INTERNATIONAL ENERGY AGENCY

RENEWABLE ENERGY POLICY IN IEA COUNTRIES

Volume II: Country Reports

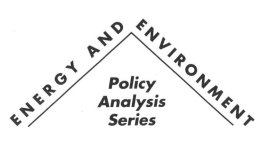

ENERGY AND ENVIRONMENT

Policy Analysis Series

INTERNATIONAL ENERGY AGENCY

9, RUE DE LA FÉDÉRATION, 75739 PARIS CEDEX 15, FRANCE

The International Energy Agency (IEA) is an autonomous body which was established in November 1974 within the framework of the Organisation for Economic Co-operation and Development (OECD) to implement an international energy programme.

It carries out a comprehensive programme of energy co-operation among twenty-four* of the OECD's twenty-nine Member countries. The basic aims of the IEA are:

- To maintain and improve systems for coping with oil supply disruptions;
- To promote rational energy policies in a global context through co-operative relations with non-Member countries, industry and international organisations;
- To operate a permanent information system on the international oil market;
- To improve the world's energy supply and demand structure by developing alternative energy sources and increasing the efficiency of energy use;
- To assist in the integration of environmental and energy policies.

IEA Member countries: Australia, Austria, Belgium, Canada, Denmark, Finland, France, Germany, Greece, Hungary, Ireland, Italy, Japan, Luxembourg, the Netherlands, New Zealand, Norway, Portugal, Spain, Sweden, Switzerland, Turkey, the United Kingdom, the United States. The Commission of the European Communities also takes part in the work of the IEA.

ORGANISATION FOR ECONOMIC CO-OPERATION AND DEVELOPMENT

Pursuant to Article 1 of the Convention signed in Paris on 14th December 1960, and which came into force on 30th September 1961, the Organisation for Economic Co-operation and Development (OECD) shall promote policies designed:

- to achieve the highest sustainable economic growth and employment and a rising standard of living in Member countries, while maintaining financial stability, and thus to contribute to the development of the world economy;
- to contribute to sound economic expansion in Member as well as non-member countries in the process of economic development; and
- to contribute to the expansion of world trade on a multilateral, non-discriminatory basis in accordance with international obligations.

The original Member countries of the OECD are Austria, Belgium, Canada, Denmark, France, Germany, Greece, Iceland, Ireland, Italy, Luxembourg, the Netherlands, Norway, Portugal, Spain, Sweden, Switzerland, Turkey, the United Kingdom and the United States. The following countries became Members subsequently through accession at the dates indicated hereafter: Japan (28th April 1964), Finland (28th January 1969), Australia (7th June 1971), New Zealand (29th May 1973), Mexico (18th May 1994), the Czech Republic (21st December 1995), Hungary (7th May 1996), Poland (22nd November 1996) and the Republic of Korea (12th December 1996). The Commission of the European Communities takes part in the work of the OECD (Article 13 of the OECD Convention).

FOREWORD

Policies used to promote renewable energy are widespread within IEA countries, and have recently been reinforced in many. Some renewable energy applications in some countries have grown significantly as a result. However, while increased environmental commitments may push governments to give further encouragement to the use of renewable energy, regulatory changes, particularly in the electricity sector, could inhibit the growth of electricity production from new renewables unless carefully handled.

This publication is the second of two volumes devoted to policy aspects of renewable energy in IEA countries. The first volume provided an overview of the policy measures used to promote renewable energy within the IEA, and an outline of the current status of renewable energy and its longer term prospects. This second volume covers recent developments in renewable energy policies and discusses the possible implications of electricity market reform for renewable energy. This book also provides more detailed information and analysis of the policies used to promote renewable energy within each IEA country, in individual country chapters.

This report is published under my responsibility as Executive Director of the IEA. It is based on the experience of IEA Member Countries, and has been reviewed in the Agency's Standing Group on Long-Term Co-operation and Policy Analysis and in the Renewable Energy Working Party. Environmental implications of increased renewable energy use are examined more fully in the IEA's recently-published "Benign Energy?:The Environmental Implications of Renewable Energy".

Robert Priddle
Executive Director

ACKNOWLEDGEMENTS

The IEA would like to acknowledge the significant assistance and support which government officials and renewable energy experts from Member countries contributed to this publication. In particular, the IEA would like to thank its Energy Advisors and delegates to the Standing Committee on Long Term Co-operation and Policy Analysis and to the Renewable Energy Working Party, who provided country-specific information on renewable energy and guidance on the report content.

This report was written in the IEA's Energy and Environment Division. The main responsibility for this work rested with Jane Ellis, under the direction of Lee Solsbery and Jean-Marie Bourdaire. Sandrine Duchesne was responsible for the statistical input. Logistical assistance was provided by Jenny Gell and Maggy Madden. Additional input, assistance and advice was provided by other past and present IEA colleagues, in particular Maria Argiri, John Cameron, Pierre-Marie Cussaguet, Stephen Peake, Karen Tréanton and Eric Savage.

TABLE OF CONTENTS

LIST OF FIGURES

ANNEX A

EXECUTIVE SUMMARY

There is a large difference in the mix of renewable energies and in the prospects for different renewable energies used within IEA countries. Biomass is the most important renewable energy source in the IEA, and all IEA countries use some form of biomass or wastes for energy purposes. The direct use of biomass in the largest single use of non-hydro renewables within the IEA, and biomass is also used in the production of heat in district heating and combined heat and power (CHP) plants. Hydropower is also an important source of electricity in many IEA countries and the most important in some. Electricity generation from other renewable energy sources has also become much more widespread in the 1990s. The use of biofuels for transport, however, is marginal, and is likely to remain so in the medium term as it is promoted in only a few countries and at a high cost. These differences in renewable energy use in IEA countries reflect different availabilities and costs, and other factors such as population density and the relative location of population centres and energy resources.

While all countries possess some indigenous renewable energy resources, harnessing them is often more expensive than using or importing a more concentrated form of energy. The cost of renewable energy, although continuing to decline, is often higher than that of other energy sources. This incremental cost has been the largest impediment to increased use of solar, wind, geothermal, and wave energy and of electricity from biomass and wastes. Moreover, many renewable energy technologies are not widely used, so they cannot yet benefit from significant economies of scale. Nevertheless, some new renewable electricity systems are now economic in niche markets (e.g. off-grid locations). The improved economics of renewable energy systems has been helped by technological breakthroughs, such as the introduction of variable speed wind turbines.

One of the most important factors favouring renewables in recent years has been political will. This is being further reinforced by the environmental commitments agreed to at the third Conference of the Parties (COP3) in 1997. A favourable policy climate will encourage entrepreneurs to initiate renewable energy schemes.

Increased use of renewable energy can have an important environmental effect. Nevertheless, few if any of the environmental externalities of energy use are incorporated into their cost, and this is one of the reasons that renewables cost more than competing energy sources.

A greater use of renewable energy can also have other positive effects, such as:

- increased energy diversity and security (through increased use of indigenous energy supplies);
- income generation from the export of new technologies;

- employment (renewable energy technologies are often labour-intensive);

- maintaining rural population levels (via incentives for biomass energy); and

- decentralising electricity supply.

Non-hydro renewable energy provided an estimated 3.9% of the IEA's energy supply, 2% of its electricity and 18% of heat production in 1996. Biomass and wastes made up the majority of renewable energy used within the IEA, and the direct use of biomass accounted for approximately half of total IEA renewable energy supply in 1996. Non-hydro renewables accounted for 3.6% of supply and 1.7% of electricity in 1990, and 15% of heat in 1992 (data unavailable for 1990).

Treatment of hydropower in this report

This publication focuses on the policies used to promote renewable energy in IEA countries. Many IEA countries are aiming to increase their use of renewable energy, often (but not exclusively) because renewable energy use produces lower emissions of greenhouse gases than other forms of energy. For most countries, the policies in force focus on encouraging increased use of biomass, wastes, geothermal, solar, wind, tidal and wave energy, and heat pumps. Small hydropower is sometimes specifically encouraged as in Portugal, but few countries promote large-scale hydro. There is, however, a potential for further hydro developments in some countries.

Water is a renewable energy that has been harnessed for many years and is widely used in some IEA countries: it supplied 2.4% of the IEA's energy in 1996, and 15.9% of the IEA's total electricity requirements in the same year. Hydropower generates more than 50% of total electricity production in Norway, New Zealand, Canada, Austria and Switzerland. For some countries, hydropower constitutes an integrated part of their renewable energy strategy.

The status of hydropower use is presented in each country chapter. However, an IEA-wide analysis of hydro policies, status and prospects is presented separately in Annex A as the policy questions surrounding large hydro development vary widely from those for other renewable energy sources. For example, hydropower systems are often large-scale; the technology is mature; the economics are favourable; expansion of hydropower is limited (as most of the best sites have already been exploited) and growth rates are lower. Statistics collected by the IEA do not distinguish between the use of large and small-scale hydro.

Note: Given the explanation above, the use of the term "renewables" in this publication excludes hydropower. This is consistent with *Renewable Energy Policies in IEA Countries, Volume I*.

The total contribution of renewables (including hydropower) to the IEA's energy supply and electricity generation in 1996 was 6.3% and 17.9% respectively.

Although the use of non-hydro renewable energy is likely to continue growing, the proportion of non-hydro renewable electricity will remain limited in the medium-term for a number of reasons including:

■ higher generation costs (except in niche markets), especially compared to gas-fired power;

■ limited resource availability for some renewables such as geothermal;

■ the intermittent nature of some renewable electricity sources;

■ increased system costs related to grid-strengthening and backup (as the proportion of renewable electricity in a system rises);

■ ancillary costs such as those related to grid-connection, infrastructure strengthening, and planning or siting, which will always be higher for smaller-scale renewable electricity systems than for larger-scale technologies.

Renewable energy policy is continuing to evolve relatively rapidly. This is due to recent changes in the energy and environment policy framework within IEA countries, and the role that increased renewable energy use can make in reducing greenhouse gas emissions. Some market-based policies, such as competitive bidding procedures within a ring-fenced market, and "green pricing" are rapidly becoming more common (especially for renewable electricity promotion). Increased information has helped some countries to learn from other countries' experiences. For example, France and Ireland both adapted schemes similar to the UK's *Non-Fossil Fuel Obligation* (NFFO). Also, some electricity distributors in Germany, Sweden, Denmark, Switzerland and the UK have followed the lead of their Dutch counterparts, who introduced "green pricing" to Europe in 1996.

Policies that have been used successfully to increase renewable electricity production in different IEA countries have included:

■ providing a guaranteed market with premium prices for renewable electricity;

■ mandating electricity purchase for a certain time period at fixed "avoided cost" levels (as in the Public Utility Regulatory Policies Act in the US[1]);

■ providing capital subsidies for renewable electricity systems and/or guaranteed markets at favourable prices (as in Spain, which has both); and

■ agreeing siting plans for wind turbines with local authorities (as in Denmark – although this is combined with other incentives including output subsidies).

---

1 However, the conditions that helped PURPA encourage renewables have since changed, notably via a drop in "avoided cost" levels.

Some of these policies subsidise the fixed costs of renewable electricity production; others provide a financial incentive for both the fixed and variable costs (although variable costs for renewable electricity generation are generally very small). However, none of these policies is completely in line with unfettered competition on price terms between electricity generators, and the effect of each policy on short-term electricity prices will be different. Capital subsidies may result, in the short-term, in a smaller distortion of electricity prices, but the results in terms of renewable electricity capacity and generation will be less predictable than those of quotas or guaranteed markets with premium prices.

Price-based competition will increase with the move towards electricity market liberalisation that is underway in many IEA countries. This changing regulatory framework renders the future for renewable electricity uncertain, and moves towards liberalisation have already affected the policies by which renewable electricity is promoted in some IEA countries. The effect of market liberalisation on renewable electricity's future depends on how the process is managed within each country.

If the overall effect of market reform on renewable electricity is negative, some IEA countries will not reach their stated targets for increasing renewable electricity generation. Emissions from the power sector could therefore be higher than expected. While consumers can express their request for renewable electricity via "green pricing" schemes, it is largely up to governments to provide an enabling environment in which renewable energy can become increasingly used, and increasingly competitive.

1

CONTEXT

Renewable energy is more environmentally benign than other forms of energy. Its use can avoid or reduce emissions of CO_2, NOx and SOx (which are emitted in smaller or greater amounts from fossil fuels). These latter gases cause acid rain, while carbon dioxide (CO_2) emissions constitute the majority of anthropogenic emissions of greenhouse gases, and therefore contribute towards global climate change. Cooling water discharge from thermal power plants may also disrupt the local aquatic environment, while nuclear power carries the very small risk of a major accident involving radioactive emissions. Large-scale hydro schemes can also have significant environmental impacts, flooding valleys and changing the level of local water tables. As well as environmental benefits, increased renewable energy use can have other benefits but as these are generally non-monetised they do not affect the relative costs of renewable and other energy sources.

The Kyoto Protocol, agreed to by 159 Parties in December 1997, sets out legally binding greenhouse gas emission limitation or reduction commitments for industrialised countries[2] for the period 2008-2012. Article 2.1(a) of this Protocol encourages some domestic polices and measures such as "promotion, research, development and increased use of new and renewable forms of energy…". Indeed, greenhouse gas emission reductions are one of the primary reasons that some IEA governments are seeking to accelerate the penetration of renewable energy.

The broad environmental and social aspects of renewable energy have encouraged many IEA governments to stimulate the use of these energy sources, even if they do not always represent the most cost-effective solution to limiting greenhouse gas emissions. But the higher cost of renewable energies combined with low fossil fuel prices and increasing competition in energy markets means that significantly increased use of renewable energy is unlikely unless governments continue to offer economic and fiscal incentives. Decreased direct government intervention and, sometimes, budgetary pressure has reduced the availability of such incentives in some IEA countries. However, increased competition within different renewable energies, such as competition between wind electricity producers, may help them to become more cost-competitive.

A detailed description of the policies used by IEA governments to promote renewable energies was given in *Renewable Energy Policy in IEA Countries, Volume I: Overview*. The present publication, *Volume II*, is divided into two parts. The first provides a summary of renewable energy use, prospects, costs, constraints and recent policy developments. The second (page 55 onwards) contains a chapter on renewable energy in each IEA country.

---

2   These are listed in Annex B of the Kyoto Protocol, and include all IEA countries except Turkey.

2

INTRODUCTION

Non-hydro renewable energies are a diverse group of decentralised energy sources that supply approximately 3.9% of the IEA's energy needs. This group of energies includes biomass, wastes[3], wind, solar, geothermal, tidal and wave. The use of some forms of energy supply from renewables is growing extremely rapidly.

There is a large variation in renewable energy use and in the absolute and relative growth rates of these energy sources within IEA countries. In most IEA countries, the largest non-hydro renewable is solid biomass (generally wood) used directly by the final consumer[4]. Biomass accounted for over 10% of total energy supply in Finland, Sweden, Austria and Turkey in 1996, and is used in all IEA countries although the majority is concentrated in just five (the US, Canada, France, Turkey and Sweden). Electricity generation from biomass and other renewable energy resources has become much more widespread over the course of the 1990s, and the use of biomass for electricity and/or heat generation is also increasing in some countries such as Sweden and Finland. The use of biofuels for transport, however, is marginal, and promoted in only a few countries, generally at high costs.

Increased renewable energy use in IEA countries is due to a mixture of technological developments, decreased costs, available incentives and a generally favourable policy climate. Since the early 1990s, IEA governments have shown increased interest in promoting renewable energy for a number of reasons, including their potential to supply greenhouse-gas free energy, and their possible contributions towards security of (energy) supply, increased energy diversity, reduced emissions of acid gases and, in some cases, maintaining rural employment. However, the environmental benefits of renewable energy are still not adequately reflected in energy pricing in the majority of IEA countries, although IEA Ministers agreed to "Shared Goals" in 1993, which include a call for "the environmental costs of energy production and use to be reflected in prices". Indeed, only a handful of northern European countries have set up a carbon tax, and gaseous emissions from other energy sources, such as SOx emissions, are explicitly valued only in the US[5].

---

3   Some wastes, such as wood waste or agricultural waste, are biomass-based fuels. Others, including municipal or industrial waste, contain a mixture of biomass and fossil-based energy, and will therefore be only partially renewable. The IEA classifies all wastes as renewable.

4   The use of biomass for energy purposes is only renewable when it is carbon-neutral, i.e. when the biomass used is replaced by new growth.

5   Although the environmental costs of SO_x and NO_x emissions are not monetised in Europe, meeting the emission reductions for these gases as set out in the Directive on Long Range Transboundary Air Pollution does of course have a cost. This cost relates to the costs of fitting equipment to reduce SOx and NOx emissions, partially addressing the environmental externalities of these emissions.

The policy packages used to promote renewables have changed over the 1990s: capital subsidies are becoming less common for larger-scale renewables (although they are still commonly used to promote rooftop PV systems), while favourable tax treatments, guaranteed markets and favourable buy-back rates for renewable electricity are increasingly used. There is also, in general, a greater emphasis on competitive bidding procedures for renewable electricity supply, information programmes and "green pricing". Competition between renewable electricity producers within a protected electricity market share is also being set up in the Netherlands as "green certificates" and in some US states as "renewable portfolio standards".

Given that there are a large number of renewable energies, each of which can be used for one or more applications (electricity generation, heating, production of secondary fuels etc.) an increasing number of governments are limiting the renewable energy sources they promote. Some countries are specifically targeting renewable electricity use in general; others are even more selective, with promotional policies directed toward one or just a few electricity-generating sources.

This selective promotion of different renewable energies for different uses is intended to maximise the short and medium-term adoption of renewable energies. Increased short and medium-term use of renewables is recognised by many IEA countries as an important part of the changes in energy supply and demand that will be needed to fulfil their commitments under the Kyoto Protocol. If, however, only commercial or near-commercial renewables (such as grid-connected wind electricity and the direct use of wood for heating) are promoted, the adoption of technologies still in the development stage would suffer; and this could ultimately limit the penetration of renewables in the IEA's longer-term energy supply. Governments therefore need to balance their expenditure on short/medium-term promotion and deployment of renewables with longer-term renewable energy goals.

3

STATUS AND PROSPECTS

The role of renewable energies varies widely within IEA countries, and between IEA and non-IEA countries[6]. For much of the population in non-IEA countries, non-commercial biomass and wastes are the only accessible energy source; biomass supplies the majority of energy in many countries in Asia and Africa[7]. However, the more modern uses of renewable energy, such as renewable electricity generation, are extremely limited. Renewable electricity systems are often more capital-intensive than traditional systems. The scope for large-scale renewable electricity technologies in developing countries is further constrained by the limited coverage of electricity grids, although this limited coverage may also increase the potential niche market for stand-alone or remote renewable energy systems.

In IEA countries, non-hydro renewables account for only a small percentage of total energy supply. In 1996, hydropower accounted for a further 2.4% of total primary energy supply (TPES). The direct use of biomass accounted for 46% of total IEA renewable energy supply in 1996. While the dominance of biomass for heating purposes is expected to remain, its percentage of total renewable electricity generation is likely to lessen in future, given the rapid pace of growth of windpower. This section examines the recent trends and the prospects to 2020 for renewable energy use in IEA countries.

RECENT TRENDS IN IEA RENEWABLE ENERGY USE

Renewable energy use in the IEA has been growing in absolute terms in recent years (figure 1) and reached 178.3 Mtoe in 1996 (equivalent to the total energy supply of Italy and Portugal in that year). The percentage of total primary energy supply (TPES) met by renewable energy sources has been stable since 1993. The fastest-growing applications of renewable energy since 1990 are in electricity production from wind, biomass and wastes, and in heat production from biomass and wastes. However, the supply of some renewable energies, such as geothermal electricity, is stable. The supply of others, such as geothermal heat, is declining.

The majority of the IEA's non-hydro renewable energy use is solid biomass: generally the direct use of wood or wood wastes. Other direct uses of renewable energy are solar and geothermal heat. In total, about half of all renewable energy

---

6 The IEA's 1998 *World Energy Outlook* includes, for the first time, projections of biomass energy use in non-IEA regions.

7 Not all biomass use in developing countries is sustainable.

supply in the IEA is used directly, i.e. without being transformed first to heat or electricity.

Solid biomass use was 121 Mtoe in 1996, 68% of the IEA's non-hydro renewable energy use in the same year. Gaseous and liquid biomass, such as biogas or transport biofuels, are also used in some IEA countries, but in smaller quantities. In 14 IEA countries in 1996, biomass supply was higher than that of any other renewable energy (including hydro). In 1996, approximately two-thirds of biomass was used directly.

At the IEA level, the largest single use of solid biomass (largely for heating purposes) is in the residential sector. This accounted for over a third of IEA biomass energy use and is an important energy source in the residential sector, especially in Turkey where it accounted for a third of energy used directly in that sector. Large quantities of solid biomass are also used in some IEA countries' industrial sectors, especially where biomass forms part of the raw materials and waste streams, as in the pulp-and-paper and wood production industries. Biomass is also being increasingly used in more modern applications such as electricity and heat generation. These uses have grown substantially in some countries, such as Sweden and Austria.

Figure 1
**Importance of Renewable Energy Supply by Country,
Showing the Relative Proportions of Biomass,
Hydro and non-Hydro Sources, 1996**

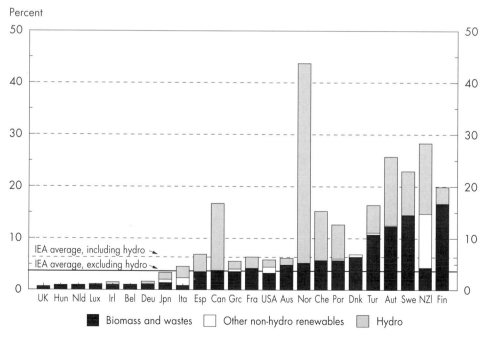

Source: IEA databases.

20

The pattern of solid biomass use varies widely among IEA countries. For example, the pulp-and-paper industry accounted for 75% of biomass energy use in Canada in 1996 and was the largest single user in Finland, Japan and Sweden. The residential sector accounted for the vast majority of biomass use in Turkey. In the US, 50% of biomass use was for generation of electricity and/or heat in 1996. Recent trends in biomass use in individual IEA countries also vary, although IEA-wide trends are difficult to determine because of the lack of data.

Differences in the use of renewable energies in different IEA countries reflect differences in energy prices, resource availabilities, and population densities, as well as in other factors. For example, easily-accessible geothermal energy is located in only a few IEA countries and geothermal electricity was generated only in seven IEA countries in 1996. Biomass energy supply also varies with population density, the relative location of population centres and forest resources. Relatively little is used in the Netherlands, which is a small, densely-populated country with limited biomass resources. It is also low in Japan, where the large, dense population centres lack easy access to biomass resources. However, it is high in Finland, New Zealand and Sweden, which are sparsely populated and have large forests, and where most wood use is in wood or paper-related industries.

Solar and geothermal energies are also used directly, as well as for electricity generation. However, the extent of the direct use of solar energy, as in individual roof-top water heaters, is extremely difficult to quantify because of its decentralised and small-scale nature. Only a few IEA countries have reported the direct use of solar and geothermal energy, so the totals shown in Table 1 are likely to underestimate the total contribution of these energies.

Non-hydro renewables generated approximately 2% of the IEA's total electricity requirements in 1996. The relative imporance of electricity production from biomass and wastes compared to other non-hydro renewables is shown in Figure 2. The proportion of electricity generated by non-hydro renewables varied from zero in Hungary to 10% in Luxembourg, where electricity from wastes is relatively important. The use of renewable electricity has grown slightly at an IEA level since 1990. However, this figure masks a wide variation in trends among countries. Non-hydro renewable electricity is increasing in most IEA countries, such as the UK and Ireland, although from a very low base in both these cases. But it has fallen off in the US (the US administration estimates that this may be a temporary phenomenon related to the restructuring of the US power sector).

Three-quarters[8] of the IEA's renewable electricity is generated from biomass and wastes (78% in 1996). This proportion has remained approximately stable

---

8   Statistics of biomass-generated electricity are difficult to quantify accurately, because biomass is often used in dual-fired stations (with coal), so reported biomass generation is necessarily calculated rather than observed. Differences in calculation methodologies, variations in the energy content of the biomass used as fuel and changes in the relative proportions of biomass and coal will all have a significant impact on the amount of biomass-generated electricity.

Figure 2
Percentage of electricity generated by non-hydro renewables⁹, 1996

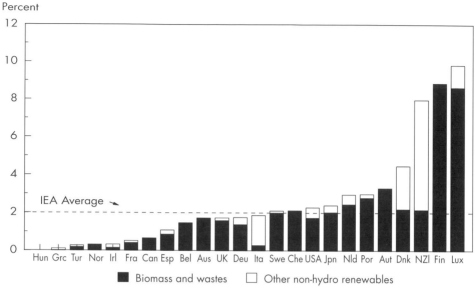

Source: IEA databases.

Figure 3
Electricity generation from non-hydro renewables, 1996

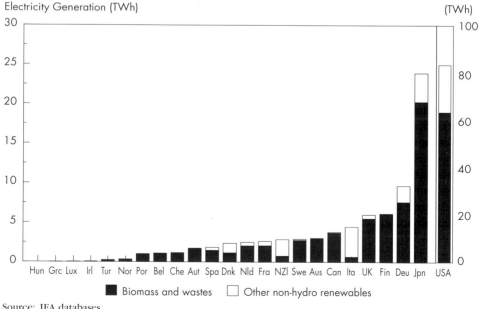

Source: IEA databases.

9   This graph refers to the percentage of electricity generated within a country that has been produced from non-hydro renewables. Luxembourg imports almost all of its electricity, but has some waste-to-energy plants: the percentage of renewable electricity generation in Luxembourg's total electricity *consumption* would be much lower.

Table 1
Renewable Energy Supply in the IEA : 1990-1996

	1990	1992	1993	1994	1995	1996
Renewable TPES* (Mtoe)	**147.2**	**160.6**	**168.8**	**171.9**	**174.2**	**178.3**
Geothermal	19.7	21.0	32.7	32.7	29.7	32.1
Wind	0.262	0.378	0.462	0.562	0.611	0.715
Solar	0.270	0.307	1.335	1.632	1.405	1.499
Tide, Wave	0.051	0.053	0.051	0.051	0.052	0.050
Combustible renewables and wastes	126.9	138.8	134.2	137.0	142.5	143.9
of which: Combustible renewables	*n.a.*	*n.a.*	*118.8*	*119.9*	*124.9*	*125.2*
of which: Wastes	*n.a.*	*n.a.*	*15.4*	*17.1*	*17.6*	*18.7*
Renewable TFC* (Mtoe)	**74.1**	**83.1**	**90.5**	**90.6**	**93.8**	**95.5**
Geothermal #	0.29	0.35	11.56	11.74	10.34	11.41
Solar #	0.209	0.238	1.252	1.555	1.325	1.408
Combustible renewables and wastes	73.6	82.5	77.7	77.3	82.1	82.7
of which: Combustible renewables	*n.a.*	*n.a.*	*75.8*	*75.7*	*81.4*	*81.5*
of which: Wastes	*n.a.*	*n.a.*	*1.9*	*1.6*	*0.7*	*1.2*
Renewable electricity (TWh)	**123.7**	**138.7**	**140.4**	**147.1**	**155.8**	**161.5**
Geothermal	23.3	24.8	25.5	25.3	23.7	25.4
Wind	3.0	4.4	5.4	6.5	7.1	8.3
Solar	0.7	0.8	1.0	0.9	0.9	1.0
Tide, Wave	0.6	0.6	0.6	0.6	0.6	0.6
Combustible renewables and wastes	96.0	108.1	108.0	113.9	123.4	126.2
Renewable heat (TJ)	**n.a.**	**190337**	**200575**	**220070**	**236455**	**266880**
Geothermal	n.a.	14389	14115	14126	14093	14080
Solar	n.a.	35	34	37	45	227
Combustible renewables and wastes	n.a.	175913	186426	205907	222317	252573
Selected Renewable Gen. Capacity (MW)*	**6085**	**7218**	**7509**	**8275**	**9282**	**9951**
Geothermal	3625	4020	4121	4226	4471	4438
Wind	1820	2553	2731	3350	4102	4801
Solar	380	385	397	439	449	452
Tide, Wave	260	260	260	260	260	260
Non-hydro renewables % TPES	3.6%	3.8%	3.9%	3.9%	3.9%	3.9%
Non-hydro renewables % of TFC	2.6%	2.9%	3.1%	3.0%	3.1%	3.0%
Non-hydro renewables % elec. generation	1.7%	1.9%	1.9%	1.9%	2.0%	2.0%
Percent of total heat generated (%)	n.a.	15.0%	14.8%	15.5%	16.5%	17.8%
Percent of total generating capacity (%)	0.37%	0.42%	0.43%	0.47%	0.52%	0.55%
Hydro electricity (TWh)	**1131**	**1150**	**1220**	**1180**	**1257**	**1291**
Percent of total electricity (%)	15.9%	15.6%	16.2%	15.3%	15.9%	15.9%
Percent of TPES (%)	2.4%	2.3%	2.4%	2.3%	2.4%	2.4%

TPES = Total Primary Energy Supply, TFC= Total Final Consumption. No estimate for biomass generating capacity is given as biomass electricity is often produced in stations co-firing coal, i.e. not all generating stations producing biomass electricity are dedicated solely to biomass. # Break in series as US data available from 1993.

Source: IEA databases (supplemented with some data from the administrations of Austria, Greece and US).

throughout the 1990s, and is consistent throughout IEA regions although it varies significantly from country to country (Figure 3). However, there have been small changes in the relative importance of renewable electricity sources since 1990, reflecting their different growth rates. For example, the relative importance of biomass and waste-generated electricity is lower than average in Italy and New Zealand, which have significant quantities of geothermal electricity; and in Denmark, where wind electricity is becoming more important.

Wind electricity has grown extremely fast in many countries over the last few years with total generation increasing from 3 TWh in 1990 to 8.3 TWh in 1996. However, the deployment of wind electricity varies enormously by region: it is negligible (although beginning to be developed) in the IEA Pacific countries; growing extremely rapidly in Europe (from 514 GWh in 1989 to over 4.8 TWh in 1996); and has stagnated recently in North America (see Figure 4). Wind energy has the advantage over other renewables that it can be harnessed for medium-scale on- or off-grid electricity generation, that it is widespread and that its availability roughly follows demand (in Europe at least, where it is windier in winter when demand is greatest). It is also cheaper than many other new renewable applications. However, it is intermittent by nature. The costs and other problems related to the integration of an intermittent resource into an electricity grid will ultimately limit the degree to which it can be harnessed.

Figure 4
Trends in wind capacity and generation by IEA region, 1990-1996

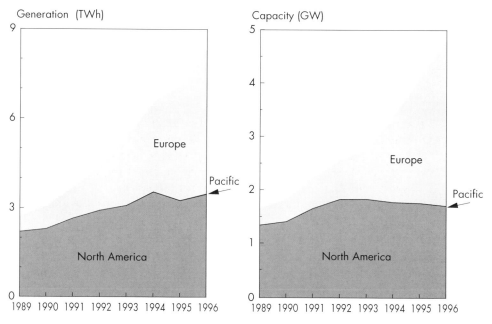

Source: IEA databases and communications with national administrations.

Non-hydro renewables are also used to produce heat: the vast majority of this was produced from biomass and wastes, although a small amount of geothermal heat was also used. The share of renewable energy in heat supply has been increasing over the 1990s. It stood at 17.8% of total heat production in 1996 compared to 15% in 1990. Although fewer than half of IEA countries use renewable energy to produce heat, it is an extremely important source of heat for some of the countries that do use it, like Switzerland. Sweden's use of biomass to produce heat in CHP systems has also grown rapidly, jumping from 24.4 PJ in 1990 to 80.8 PJ in 1996.

PROJECTED RENEWABLE ENERGY SUPPLY IN THE IEA

Projected renewable energy supply in the IEA is expected to increase substantially to 2020, both in absolute and proportional terms. The contribution of biomass and wastes is projected to grow to 172 Mtoe in 2020 from 148 Mtoe in 1996[10]. Total energy harnessed from geothermal, wind, solar, and tide is projected to stand at a combined 54 Mtoe in 2020, or 0.9% of Total Primary Energy Supply (up from 0.7% of TPES in 1996). Renewable electricity from these sources is projected to generate 153 TWh or 1.2% of the IEA's electricity in 2020 compared to 37 TWh and 0.4 % in 1996.

The latest version of the IEA's *World Energy Outlook* (WEO)[11] projects in the "business-as-usual" case that fossil fuels will account for the vast majority of the IEA's total energy supply in 2020 (see Table 2). Nevertheless, it is more optimistic than before on the outlook for renewable energy in Europe and the Pacific, particularly for renewable electricity, which it projects to grow at an annual average of 12% to 2010 (see Table 3). The majority of this growth is expected to come from geothermal electricity in the Pacific and from wind and biomass in Europe.

However, the rate of growth of renewable energy in general, and of renewable electricity in particular, is projected to vary significantly among IEA regions (Table 3). Some of the largest variations are expected in wind electricity: North America is expected to grow only slowly from a relatively high base[12], Europe is expected to grow rapidly from a high base, and the Pacific is projected to grow rapidly from a low base. The relative changes for different renewable electricity sources are illustrated in Figure 5.

---

10   Projections are available for a slightly different group of countries than those that make up the IEA: Czech Republic and Iceland are included. The 1996 supply figure is therefore different from that quoted in Table 1.

11   IEA, 1998.

12   This is a significant downwards revision from WEO96, and reflects US uncertainty surrounding renewable energy development in a deregulated electricity market.

Table 2
Projected energy supply by fuel type, OECD Total*

	TPES (Mtoe)			Electricity generation (TWh)		
	1996	2010**	2020**	1996	2010	2020
Solids	1148	1256	1391	3420	4053	4777
of which Biomass and Wastes	*148*	*159*	*172*	*127*	*143*	*164*
Oil	1912	2159	2262	531	551	585
Gas	994	1329	1433	1031	2872	4041
Nuclear	524	516	438	2011	1978	1681
Hydro	112	121	128	1300	1410	1483
Geo/Other	35****	42	54	37	93	153
Total*	**4726**	**5423**	**5707**	**8329**	**10957**	**12721**

* OECD excluding Mexico, South Korea and Poland.

** Projections are based on official IEA data and may exclude direct use of geothermal and solar heat in some countries if these data are unavailable for 1996.

*** Totals may not add, due to rounding and to exclusion of electricity trade from TPES figures.

**** Including US national data.

Source: IEA *World Energy Outlook*, 1998. "Geo/other" figures are IEA (Energy and Environment Division) Secretariat estimates.

Table 3
Projected growth rates for energy and electricity from new renewables by OECD region*

	TPES** (Mtoe)			Electricity generation** (GWh)		
	1996	2010***	% p.a. growth	1996	2010	% p.a. growth
N. America	26****	18	–	20	29	2.6
Pacific	5	13	6.8	6	18	8.4
Europe	5	11	6.6	10	46	11.8
Total OECD	**35**	**42**	**–**	**36**	**93**	**7.1**

* Includes geothermal, wind, solar, tide and wave.

** Excludes Mexico, South Korea and Poland.

*** Projections are based on official IEA data and may exclude direct use of geothermal and solar heat in some countries if these data are unavailable for 1996.

**** Includes US national data for 1996 only (which is 11 Mtoe higher than that reported to the IEA due to the inclusion of the direct use of solar and geothermal heat).

Source: IEA (Energy and Environment Division) Secretariat estimates, 1998.

Figure 5
Current and projected non-hydro renewable electricity generation in the OECD, by source, 1996, 2010, 2020

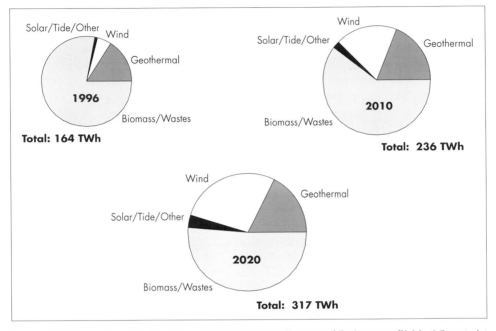

Source: 1996 figures from IEA databases, projections are IEA (Energy and Environment Division) Secretariat estimates, 1998.

Other studies also project the use of renewable energies to grow[13]. For example, the European Commission study "The European Renewable Energy Study II" contains a number of different scenarios that indicate a growth in renewable energy (including hydro) from just over 4.5% of TPES in 1996 to between 7.4% and 14% of TPES by 2020. Official 1998 US forecasts also show renewables growing, although slowly because of their cost disadvantage in a deregulated market. The US' 1997 forecast projected a significantly higher growth.

---

13   See also *Key Issues in Developing Renewables*, IEA, 1997.

RENEWABLE ENERGY BENEFITS

Increased use of renewable energy that displaces or delays increased use of other fuels can help reduce the negative environmental impacts associated with these fuels[14]. This is often the main reason for which renewable energy is promoted. In addition, the dispersed nature of renewable energies means that all countries possess some indigenous renewable energy resources (to a greater or lesser extent); harnessing those energy sources can therefore help improve energy security.

Quantifying the environmental benefits of increased renewable electricity is difficult, and doing so in *monetary* terms is even more complex and uncertain, since the value of avoided emissions is subject to disagreement, and may even vary with time and location. (For example, the damage caused by acid rain will depend on the height at which SO_2/NOx emissions were released, how far these gases were transported, the rainfall pattern along the route, and the area on which it is deposited).

A handful of northern European IEA countries have attempted to address some of the environmental externalities associated with energy use by putting in place a carbon tax. Such a tax places an explicit value on CO_2 emissions, and hence on the environmental impact of such emissions. Even in these cases, however, fossil fuel inputs to electricity generation and other bulk uses of fossil fuels are wholly or partially exempt, so the impact of this measure is limited. Nevertheless, the introduction of a carbon tax is credited with dramatically increasing the use of biomass for heat generation in Sweden. A number of US States have used monetised values for some environmental externalities in their resource planning process. But these values were not always at a high enough level to affect significantly the outcome of planning decisions. Auctionable permits for SO_2 emissions were set up by the US as part of the procedure to reduce gaseous emissions under the Clean Air Act Amendments. But the availability of low-sulphur coal (and the ability to export high-sulphur coal to Europe) have resulted in utilities continually over-complying with these regulations, and steep reductions in the price of SO_2 emissions permits[15].

Since environmental and other benefits of renewable energy are not systematically taken into account in fuel choice decisions, competition between renewable

---

14 Of course, there are also negative environmental impacts of renewable energy use, especially if examined over the full life-cycle of production and decommissioning, although the disbenefits are generally small in comparison to renewable energy's benefits. The environmental impacts of renewable energy use are discussed in the IEA publication: *Benign Energy?: The Environmental Implications of Renewables* (1998).

15 SO_2 emissions permits cost around \$100/t S, compared to a penalty level of around \$2,000/t S.

energy and other energies is not on an equal footing. Although some IEA countries are working towards an increased monetary valuation of the environmental benefits of benign energy sources, and although the IEA's Shared Goals call for countries to reflect environmental costs in energy prices, this is not yet common practice, and is unlikely to become so in the near future.

RENEWABLE ENERGY COSTS

Increased use of renewable energy entails costs as well as benefits. In most cases, the cost of renewable energy is higher than that of other energy sources, although extremely site-specific. This incremental cost has been the largest impediment to increased use of renewable electricity in IEA countries. Cost comparisons for selected IEA countries are given in Table 4.

The incremental cost of many renewable electricity sources is partly due to the dispersed nature of renewable energies, and partly because many renewable energy technologies (e.g. solar thermal electricity production) are not yet mature, and sales are not yet at a high enough level to benefit from significant economies

Table 4
**Comparitive costs of renewable and other electricity,
selected IEA countries (cents/kWh)**

Country	Projected costs of new baseload plants#			Renewable electricity costs		
	Coal	Gas	Nuclear	Wind	Biomass	Other renewable
Canada	3.7-5.4	3.3	4.0-4.7			33.2* (non-grid PV)
Denmark	4.9	5.2-5.8	-	5.5 (on-shore)	11.9 (straw)	
France	6.0	5.3	4.9	6.5**		
Germany	n.a.	3.3	n.a.	5.8-11		11-14.4 (Hydro <0.5MW)
Italy	5.3	5.1	-	7.5		
Japan	7.6	8.4	8.0	13.3*		
New Zealand***	5.0	2.8	-	4.5-6.3		3.6-5.7 (geothermal)
Netherlands	5.6-6.2	4.4-4.8	-	9.0* (1995)		
UK (England and Wales)		3.9*		5.5**	5.1** (MSW)	
US	3.5-3.6	2.4-2.7	4.6		4.2 (biomass)	

\#   (10% discount rate, 1996 $US) Source: *Projected Costs of Generating Electricity*, IEA/NEA, forthcoming 1998.

*   1997 cost estimates from National Administrations

**   Average bid prices for successful projects, 1997

***   New Zealand Energy Outlook, Ministry of Commerce, February 1997 of scale. Moreover, no monetary values are given to many of the environmental and social benefits that increased use of renewables can bring, so these benefits are ignored in purely economic comparisons of renewable energies with other energy sources. In addition, renewable energy sources are in the unenviable position of having to compete with established energy sources that can benefit or have previously benefited from advantages such as subsidies, available infrastructure and fully amortised capital investments.

Increased electricity generation from combined-cycle gas turbines, together with persistently low fossil fuel prices, has brought declines in baseload electricity prices in many IEA countries over recent years. Nevertheless, the incremental cost differential between renewable electricity and other electricity has also lessened in some cases, as the cost of some renewable electricity sources, particularly wind, has dropped rapidly (Table 5). This cost reduction includes the effect of technology learning[16]. Some renewables are now economic in niche markets such as remote locations not connected to a main electricity grid, locations near a supply of biomass energy, or in areas where using waste streams as an energy source reduces waste disposal costs[17]. However, there may be additional costs associated with introducing intermittent renewable electricity generation, such as the cost of additional back-up capacity and grid-strengthening.

These cost improvements have been helped over recent years by a number of policy, technological and economic factors. One of the more important, although least tangible, has been a greater political will to promote renewables over a period of time: many IEA countries have been actively promoting renewables since the Earth Summit in 1992 or even before. A favourable and stable policy climate, backed up by economic and fiscal incentives, encourages entrepreneurs to initiate renewable energy schemes, and indirectly facilitates the finance of such systems by raising financiers' confidence in renewable energy systems.

Another important factor in cost reduction for some renewable systems, especially wind electricity, is the scaling up of the technologies involved and the consequent reduction in unit construction, connection and operating costs. The capacity of newer wind turbines installed can be as high as 1.5 MW, compared to 300 kW a decade ago[18]. For wind energy systems, the development of variable-speed turbines is also important. In addition, the competitive bidding procedures initiated by

---

16 This learning is quantified in "experience curves", which are used as a tool for strategic planning in many high-tech industries. A detailed discussion will be available in *Technology Learning and Future Technology Availability* (IEA, forthcoming 1999).

17 This situation could arise either in certain industries, like pulp and paper, where using biomass residues reduces waste disposal costs. It could also arise in densely populated countries, where land for landfills is scarce, and the opportunity costs of using this land is high.

18 The rate of technology development may in itself lead to some technology or economic risk. For example, since there is only a small body of any one turbine type in operation, long-term stresses or flaws may only become noticeable a few years into the future. Moreover, the recent bankruptcy of the US Kenetech and German Tacke wind manufacturers means that technology procurement is not without risk.

Table 5
Cost trends for wind electricity, selected IEA countries (US cents/kWh)

	1991	1992	1994	1995	1996	1997
England & Wales	18		7.1			5.5
Scotland			6.4			4.5
Denmark						5.5
Netherlands	9.7	8.8	7.1	6.9	6.4	5.3
Norway		5.0				4.2

Source: National administrations and their communications to the IEA.

some countries whereby developers of similar renewable energies compete on cost grounds amongst themselves. have also helped to push costs down for renewable electricity generation. These include the UK's *Non-Fossil Fuel Obligation* or Ireland's *Alternative Energy Requirement.* Increased availability of finance at lower interest rates is indispensable in lowering the cost of these capital-intensive technologies. This can be achieved where financiers are willing to lend at lower interest rates to technologies they perceive as more developed and less risky; or through larger companies self-financing renewable electricity projects rather than borrowing.

The costs of encouraging renewable energy development includes the direct costs of incentives, such as capital costs, output subsidies, tax rebates, and information/education campaigns. It also includes indirect costs, such as that of implementing legislation, standards or other regulations that promote the use of renewables. Other costs associated with increased use of intermittent renewable electricity relate directly to the electricity system itself, such those for strengthening the transmission and distribution system and of providing backup capacity. These issues are discussed below.

DIRECT COSTS FOR RENEWABLE ELECTRICITY PROMOTION

The costs of government incentives are generally passed on to the consumer through higher electricity prices and tariffs. This is done either directly, as in the UK's levy on electricity prices, or indirectly, as in Germany, where the cost of supporting renewable electricity is placed directly on the utilities, who can then pass it on the consumer.

SYSTEM COSTS RELATED TO RENEWABLE ELECTRICITY SUPPLY

Since electricity cannot be stored effectively on a large scale, generation and demand have to be matched. However, demand fluctuates constantly. Although the general shape of 24-hour load patterns is well known by season and country, day-to-day fluctuations can be significant. In order to avoid brownouts and blackouts, the electricity system has to be able to satisfy instantaneous changes in demand.

Matching supply and demand is done by ensuring that there is always capacity that can be brought on line either to satisfy changes in electricity demand, or to replace a power station if it fails. This capacity falls into four types:

- "spinning reserve", capacity which can provide power instantaneously on demand;

- load-following capability, which can provide short-term power increases, in about an hour;

- backup capacity, which can provide power within a couple of hours or more; and

- mothballed capacity, which can take many days to bring back into service.

In addition, electricity imports may be used to enable countries to meet unexpected levels of demand, or partially to offset plant failures and/or shutdowns.

Maintaining some reserve capacity is necessary to ensure reliability of electricity supply, but it is also costly. If electricity generation from intermittent renewables, such as wind and solar, increases to such an extent that the consequent variations in electricity generation require an increased reserve margin, the system's total backup costs will increase. However, the current use of wind and solar is too small for increased backup costs to be an issue. Nevertheless, renewable electricity may become more important, especially in Denmark, if the ambitious aims for renewable energy supply as outlined in their *Energy 21* plan are achieved[19].

Transmission issues are also important in ensuring the reliability of electricity supply. Electricity has to be transmitted from its supply source to the point of demand. Since a small amount of electricity is lost during transmission, transmission over longer distances will be more costly than over smaller ones. While smaller-scale renewable technologies, such as rooftop photovoltaic systems, produce power at the point of demand and therefore have low or no transmission costs, larger-scale renewable electricity systems, such as wind turbines, are often located far away from demand centres.

---

19   The Danes, however, do not perceive this as a problem, given that there is significant trading capacity between Denmark and its neighbours and that the wind sites to be developed are offshore (where it is usually more windy than on land).

The strength of the transmission grid is also important. Within any one grid, transmission lines are generally interconnected so that if one line goes down, power supply can be maintained by an alternative route. Therefore, there are usually many interconnections between higher voltage level power lines, although fewer at the extensive, low-voltage levels (i.e. at the level of electricity distribution). This makes the lower-voltage grid weaker. Generation capacity that is widely distributed and often in remote or rural areas, e.g. from wind or biomass, may be nearer a low-voltage than a high-voltage grid. Weak low-voltage grids are unlikely to be able to support the connection of more than a few MW of generation capacity (particularly from an intermittent source) without being strengthened. Strengthening the grid, which entails the installation of new or upgraded power lines, is costly. The issue of who pays for any grid-strengthening may affect the rate of adoption of renewable electricity technologies. (The current situation in IEA countries varies – for wind electricity, it is the utility who bears the cost of grid-strengthening in Denmark, whereas it is the turbine operator who pays in Germany).

The cost of connecting renewable electricity systems to the electricity grid may also be high, especially for wind electricity, where good resources are often located in remote, rugged areas (or off-shore), far from an existing electricity grid. Where grid-connection is expensive, determining whether the independent generator, utility, transmission authority or distributor pays for it will also be a contentious issue. Expensive connection costs may mean that a large but remote wind resource may only be economic if it is developed in large capacity blocks, rather than building slowly. However, this will also increase investment costs, and could inhibit the ultimate development of such sites, particularly in liberalised markets where the risk of not being able to sell electricity makes generators aim to minimise their capital exposure.

BILLING PROCEDURES

The generation, transmission and distribution functions of electricity supply are becoming increasingly separated in a deregulated market. It is likely that the cost of electricity to consumers will also become more "unbundled", and that in the future, the billing of consumers and of generators will reflect their marginal costs.

Generators that produce a steady load throughout the year, and consumers who demand a steady load throughout the year, will face lower costs for backup capacity than generators and consumers who have highly variable generation and demand. These lower costs will be reflected in electricity bills or electricity payments for generators in a system where billing includes not only the costs of electricity generated, but also the cost of transmitting that electricity to the consumer, and the cost of system backup. If such a billing system comes into operation, it will reduce the number of potential wind farm sites, and therefore the potential installed capacity. It may even affect the way in which turbine operators choose to configure their systems (see box), and may actually result in *lower* quantities of wind electricity being generated per turbine, although with a lower variability.

Wind electricity variations

Wind energy operators try to balance the costs and benefits of variable speed turbine operation to maximise profit. Finding the most efficient balance is difficult, as it depends on many factors.

First, variations in wind speed will affect electricity output. When there is no wind, or when wind speeds are very low (<A, often 3-4 metres/s), the turbine will not generate any power. Electricity is only generated between the wind speeds A and C. At very high wind speeds, (around 25 m/s) the turbine will shut down for safety reasons. Maximum power is generated at speed B (usually around 7-10 m/s). A small drop in wind speed can result in large drops in power output because this is proportional to the cube of the wind speed. Halving the wind speed results in the power dropping by a factor of 8. This is illustrated in Box

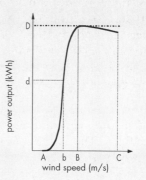

Box figure 1: power output variability

figure 1, where maximum power, D, is produced at speed B. A small drop in wind speed from B to b results in a significant drop in power output, from D to d.

The value of B has a huge impact on the power produced from a turbine. The level chosen for B will depend on the pattern of wind speeds at a site throughout the year. The exact shape of the curve in Box figure 2 will vary by wind site, with the average wind speed moving closer to the mode wind speed if very strong winds are rare, or further from the modal wind speed if strong winds are more common. Since the power potential varies with the cube of the wind speed, power generation will be maximised by making B larger than the average wind speed.

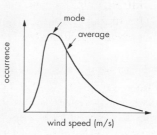

Box figure 2: wind speed distribution

Second, varying the exact configuration of a wind turbine will alter its cost, and therefore the cost of the wind electricity it produces. More power can be produced at greater heights, with a larger rotor and/or a larger generator. However, increasing the height of a tower, the length of the blades or the size of the generator is also costly.

These factors have to be balanced when determining a wind turbine's configuration. Although wind speed *patterns* may be known, accurate forecasts cannot be made for wind speeds (or power outputs) more than a few hours in advance. Therefore, the power output from a turbine is unpredictable, as well as intermittent.

6

CONSTRAINTS ON RENEWABLE ELECTRICITY USE

At present, electricity generation from non-hydro renewables is fairly limited in most IEA countries: renewable energies generated 2.0% of the IEA's total electricity in 1996. Apart from Luxembourg, Finland has more renewable electricity than any other IEA country, 9.6% of its total electricity production. The penetration of renewable electricity is likely to remain limited in the medium term in many IEA countries, as it is constrained by factors other than its incremental cost and (in some cases) limited resource availability. These constraints include:

■ the intermittent nature of some renewable electricity sources, which means that they can only be used for a limited proportion of a system's electricity supply. There is no consensus as to what exactly this limit is, with estimations varying between 5% and 25% of total supply;

■ higher system costs related to grid-strengthening and backup as the proportion of renewable electricity in a system rises;

■ the small-scale nature of many renewable electricity technologies (especially PV, but to a lesser extent wind). The dispersed nature of renewable energy source means that ancillary costs such as those related to grid-connection, infrastructure strengthening (e.g. for roads), and planning and siting will always be higher per unit of output than for larger-scale technologies. For biomass systems, fuel transport cost is especially important.

The fastest growing renewable energy application in IEA countries is wind energy, which grew in Europe at an average 15.7% p.a. between 1990-1996. Wind electricity provided approximately 2.3% of Denmark's electricity in 1995, and more than 3% in 1996 – more than any other country. Most turbines to date have been installed on land, but in Denmark, and some other northern European countries, attention is now being directed towards offshore power. It is windier offshore and more consistently windy than on land and the development of offshore power would not be subject to the siting problems that have delayed or even stopped plans for some on-shore wind farms.

Power output from all wind energy systems is intermittent (see Box above), and the need for system reliability means that wind energy can only be relied upon to generate up to a certain proportion of a system's electricity supply. However, since the proportion of wind energy has been extremely low in all systems until very recently, the costs and difficulties of managing intermittent loads have not been accurately quantified. So the upper limit on intermittent electricity generation remains uncertain. This limit will also be affected by just how intermittent the wind (or solar) electricity source is. This in turn will depend not only on the location and distribution of renewable electricity systems, but on the configuration of these systems.

Since the incremental cost of renewable energy is the largest impediment to its becoming a more significant electricity supplier in IEA countries, governmental backing for it often takes the form of economic and fiscal incentives: capital subsidies, soft loans or reduced VAT rates. However, other policies are also used to promote renewables. These include: guaranteed markets for renewable electricity, often at favourable prices; regulatory measures and standards; information and education activities; renewable energy and/or electricity targets; voluntary agreements; "green pricing" (a company/industry initiative) and R&D activities. These policies are described in detail in *Renewable Energy Policy in IEA Countries, Volume I: Overview*. Recent policy developments are discussed in a section below.

7

RENEWABLES IN A LIBERALISED ELECTRICITY MARKET

The policies that many countries have put in place to promote renewable energy emphasise the promotion of renewable electricity. This is partly because of the large potential for renewable electricity supply and partly because an increasing proportion of this potential is becoming economic.

The structure of the electricity industry is changing rapidly in many IEA countries. The presence of large, vertically-integrated utilities with a guaranteed customer base is becoming less and less common as governments liberalise their electricity sectors. While liberalisation takes many forms, "unbundling" – the separation of generation, distribution and supply functions – is becoming more and more widespread, as is competition among electricity generators to supply electricity.

In a competitive market, generators bid for the price at which the central pool/exchange will take their electricity, with large users being able to generate their own electricity or negotiate directly for supply from a generator. Competition also applies to distribution, with local distribution companies and even end-users increasingly able to choose their electricity supplier. For example, Swedish, Norwegian and Finnish customers can choose their supplier in the deregulated Nordic electricity market, and competition is also opening up more choice to customers in the UK and US. Customer choice of electricity supplier will continue to expand over the next few years with the opening of European electricity markets to competition.

Since both renewable electricity technologies and the market in which they are operating are evolving, so is the manner in which renewable electricity generation is encouraged. Some IEA countries (whose electricity industries are under very different regulatory regimes) are beginning to use market-focused mechanisms to promote renewable electricity: the UK's *Non-Fossil Fuel Obligation*, Ireland's *Alternative Energy Requirement* and France's *EOLE* programme all create guaranteed electricity markets at favourable prices[20] for a selected number of government-chosen projects, although renewable electricity projects aiming to supply this guaranteed market must compete amongst themselves for the privilege.

Two IEA countries have recently initiated another approach to promoting renewable electricity by guaranteeing it a market *share* rather than a guaranteed market for the output from selected renewable electricity systems. Under the Netherlands' system of "green certificates", which was outlined in their *1998 Electricity Act*, the government will set from 2001 the percentage of electricity

---

20 For these policies, the actual price paid for renewable electricity to different projects varies, and reflects the price bid by the developer (assumed here to be at such a level as to be profitable). All these policies are described in more detail in the respective country chapters.

supply that is to come from "sustainable sources". The result will be a guaranteed market share for renewable electricity. US proposals for a "renewables portfolio standard" are similar, and are in the process of being implemented in some US states.

The policies outlined above presuppose an element of competition between renewable electricity suppliers, thereby ensuring that potential projects have to compete against similar projects for a guaranteed market. This helps drive down renewable electricity costs, and therefore the financial support required per kWh. The provision of incentives for renewable electricity is not completely in line with a free electricity market. However, it helps renewable energy surmount the barriers caused by the non-inclusion of environment externalities in energy pricing and by continued subsidies or other incentives to other fuels, especially domestic coal or nuclear power.

Countries that have relied predominantly on market mechanisms to promote renewable electricity have witnessed only a very small expansion of these technologies. This is illustrated in France, where, although the state utility was obliged to purchase any renewable electricity offered to it, the rates offered were so low that there was little interest in new renewable generation until the start of the EOLE programme. Another example is the low penetration of PV systems in remote areas of Australia where the emphasis of PV promotion is on increased availability of information and of some low-interest loans. Hence, despite the presence of a significant niche market, i.e. houses not connected to the electricity grid in areas of high solar insolation, the penetration of PV systems is very small.

ARE ELECTRICITY MARKET LIBERALISATION, INCREASED RENEWABLE ELECTRICITY AND GREENHOUSE GAS COMMITMENTS IN CONFLICT?

The Protocol agreed to at COP3 in Kyoto in December 1997 has not yet entered into force, and will not do so until at least the turn of the century. However, if the Protocol does come into force (which is not yet certain) Parties will need to establish plans to meet their commitments to reduce greenhouse gas emissions, as these commitments will be legally binding.

It is likely that the energy sector will be a focus for many countries' emission reduction policies, as the energy sector is responsible for the majority of anthropogenic greenhouse gas emissions in IEA countries. Since there are many low-cost options for reducing emissions from the electricity sector, containing emissions from the electricity sector is also likely to feature strongly in many IEA countries' implementation of their commitments.

The carbon intensity of electricity production has been falling steadily over time in IEA countries, due both to changes in fuel mix, including increased use of nuclear

Figure 6

**Variations in the carbon intensity of electricity production,
selected IEA countries**

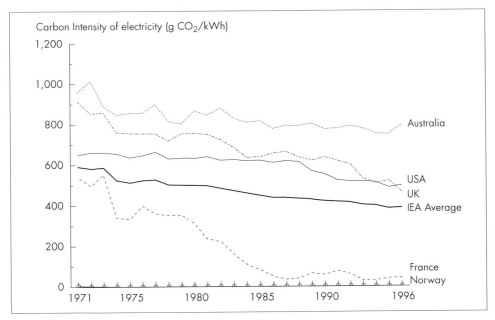

Source: IEA databases.

power, and to increased generation efficiency, notably the development of combined cycle gas turbines. Figure 6 illustrates this trend, and also indicates the variation in different countries' carbon intensity. Despite this reduction in carbon intensity, the proportion of total energy-related CO_2 emissions from electricity generation continues to grow because of the trend toward an increased share of electricity use in total energy use.

Increased competition is expected to drive down electricity prices as most of the competition between electricity suppliers is focussed narrowly on tariffs. However, other factors such as energy management, variable billing schemes or any other activity which may increase "brand loyalty" may also be factors in deciding whether or not customers will change electricity supplier. In a competitive market, utilities will be more unwilling to bear costs that will make their electricity more expensive than that of competitors. Although some consumers have stated a preference for "green electricity"(some are actually participating in green pricing schemes) it would be unrealistic to suppose that most consumers, given the option, will not choose the cheapest power on offer, irrespective of its environmental characteristics. Indeed, the limited experience of consumer choice to date shows that customers may switch electricity suppliers in significant proportions – up to 25% in the Nordic market. Less movement, however, has been noted in US markets, particularly California. Some consumers may even switch from using electricity to using other fuels during periods of high electricity prices.

Working for an increased use of renewable electricity and a fully liberalised electricity market will not be straightforward. Despite some impressive cost reductions over recent years, renewable electricity generating costs are higher than those from other sources. Therefore, increased competition on cost grounds alone still is likely to disfavour renewable electricity, except in niche markets and where customers are prepared to pay a premium for green electricity. Hence, governments wishing to increase the importance of renewable electricity will have to continue supporting it either until it becomes completely cost-competitive, or until the negative externalities of other electricity-generating technologies are included in their cost. Moreover, a move towards electricity market liberalisation may result in countries' removing or weakening policies that had been introduced to promote renewables.

This can be illustrated by recent changes in Finland. The Finns had set up a carbon tax which was applied to inputs to electricity generation. However, the liberalisation of the Nordic electricity market meant that costs facing Finnish generators were much higher than those facing their Norwegian and Swedish competitors (who rely almost exclusively on carbon-free electricity sources). In order to improve the competitiveness of Finnish generators in the NordPool/El-Ex electricity exchanges, the carbon tax was replaced by a tax on electricity production (although the carbon tax remains in force for fuel inputs to heat production). This tax is applied equally to renewable-based and other electricity, since they cannot be distinguished at the point of consumption.

Another example is seen in Italy, where the Directive mandating favourable prices for renewable electricity and for other independently-generated electricity for the first eight years of a plant's operation (CIP 6/1992) was rescinded. This action was taken as part of the government's recent reorganisation of the electricity sector and the subsequent abolition of "overpricing" for electricity from certain sources, including renewables. Removing these favourable prices will stifle the interest that had recently been shown in Italy in renewable electricity unless other (economic) incentives are introduced in their place. Germany is also changing its legislation on renewable electricity, recognising that the Electricity Feed Law (EFL) places an uneven burden on different utilities, and so affects their relative competitiveness. Under the EFL, German utilities have to buy any renewable electricity offered for sale by independent generators, and at premium rates. This has placed a heavy burden on some of the utilities nearest the coast, where wind resources are high. Legislation currently going through parliament would cap at 5% the amount of electricity that any local or supra-regional utility has to distribute[21], and so may ultimately cap the amount of renewable electricity generated. This cap is not expected to result in any renewable electricity being refused until at least 2000, by which time the legislation may have changed again.

The question is, then, how will governments balance their increasingly hands-off intervention in the electricity sector with the aim of increasing renewable

---

21 More details of this proposed legislation are given in the German country chapter.

electricity and limiting greenhouse gas emissions from electricity generation? There are a number of options:

- continuing to pursue both liberalisation and increased renewable electricity use. Such a strategy recognises that protecting renewables to a certain extent from full competition on short-term cost grounds will increase their use, but requires funds to do so;

- introducing full-cost energy pricing by the incorporation of externalities into the cost of energy. This will have the effect of raising the price of fossil fuel and nuclear electricity compared to renewables, although it may not be a politically acceptable solution in most IEA countries; and

- domestic emissions trading. The role that this could play is as yet uncertain, as national emission targets have not yet been allocated amongst sectors and users, and the rules and modalities for emissions trading have not yet been set up.

Although electricity market liberalisation is likely to affect the growth of renewable energy, the full impact is not yet clear, and many questions are as yet unanswered. Will liberalisation throw into question the concept of protected markets that have succeeded in increasing the importance of renewable electricity in the UK and are beginning to do so in other countries? Granting that the cost of renewable electricity is likely to continue to decline, will the differential between the costs for renewable electricity and fossil-generated electricity also narrow? Will low fossil fuel prices slow the progress of renewable electricity's market share? Governments should be careful that the liberalisation of electricity markets is not carried out in such a manner that it stifles the renewable electricity technologies which are beginning to take off.

8

RENEWABLE ENERGY POLICY DEVELOPMENTS

Volume I of *Renewable Energy Policy in IEA Countries* discussed the policy measures used to promote renewable energies in these countries. This section of this publication summarises that discussion, indicates developments in renewable energy policies that have occurred since the publication of Volume I and outlines situations in some IEA countries where energy policies may inhibit the uptake of renewable energies.

Renewable energy supply, and renewable energy policy developments have been evolving at different paces in different IEA countries. Some countries have a longer history of using "new" renewable energies than others, and some IEA countries have been promoting new renewable energies more vigorously or for a longer time than other countries. For example, Japan and Italy both have a long tradition of using geothermal electricity; Finland generates a relatively large proportion of its electricity from biomass; Denmark, with almost no indigenous coal reserves and most of its electricity needs being met by coal-fired power stations, has been actively promoting renewable electricity use for over a decade. California had a well-documented 'wind rush' in the 1980s although this stagnated to a large extent at the beginning of the 1990s.

On the other hand, Norway, whose electricity demand is met almost exclusively from hydropower, has had little need to develop new renewable electricity capacity. Interest in new renewable electricity has, however, increased since the high electricity demand and extremely low rainfall in 1996 combined to turn Norway from a net electricity exporter to a net importer that year. Australia, with its plentiful reserves of coal, oil and gas, and France, with its large nuclear park, have also concentrated energy expenditure (both R&D and incentives) on non-renewable energy resources.

The policies that IEA countries have introduced to promote renewable energy may be designed to overcome:

- market barriers, such as lack of information;

- some market distortions, especially the non-monetisation of the positive environmental benefits of renewable energy;

- technical barriers, by creating a market for and experience with a technology type;

- cost barriers.

Alternatively, they may be designed to create an institutional framework that is readier to accept power from decentralised sources.

Renewable energy policy is continuing to evolve due to recent changes in the policy environment (such as electricity liberalisation and the greenhouse gas commitments agreed to at Kyoto) and to technological and cost improvements for the different renewable energies.

Economic and fiscal incentives are the most widely used measures to promote renewable energy in IEA countries. These are directed at a wide range of renewable energies and renewable energy applications (direct use, heat production and electricity generation). It is these measures that have been most successful in promoting renewable energy supply – irrespective of whether this has been done in a market-based manner or not. This is because a large proportion of those who buy renewable energy equipment are private firms or individuals, susceptible to economic arguments: for example, independent renewable electricity producers are in the business of electricity generation to make a profit.

Many IEA countries have introduced policies that explicitly aim to increase the supply of renewable electricity. Effective promotion of renewable electricity will be facilitated if it is examined in the context of the *whole* electricity market. Market reforms in the electricity sector in many IEA countries are influencing the policies used to promote renewable electricity. These new policies increasingly contain an element of intra-renewables competition.

Competition among different potential fuel inputs for electricity generation is uneven on a purely economic level. This uneven competition arises in different

Possible means of using market mechanisms to promote renewable electricity include:

☐ ensuring independent power producers and non-utilities have access to the electricity grid;

☐ allowing distributed generators to feed into and take from the grid;

☐ setting aside a (small) protected and guaranteed market, even a very small one, for renewable electricity*;

☐ creating supply-side incentives, such as favourable buy-back rates, to make renewable electricity generation economically attractive for potential generators*;

☐ creating a market for renewable electricity by requiring a certain proportion of total electricity to come from renewable sources; and

☐ making demand-side incentives such as "green pricing" as widespread as possible.

* Temporary measures

ways for different reasons. Examples of policies that have favoured renewables include:

■ encouraging increased renewable energy via generous economic and fiscal incentives as in Spain, where capital and output subsidies are available for renewable electricity systems; and

■ creating a protected market for limited quantities of renewable electricity, as in the UK, France and Ireland.

Other, non-renewable, fuels have also enjoyed the benefit of explicit encouragement for their increased use. For example, the "fossil fuel levy" on consumers' electricity bills in the UK was initially used to subsidise nuclear power (as well as renewables), and the "Jahrehundertvertrag" was used to require German utilities to buy German coal. The 1997 Spanish electricity act allows up to 15% of electricity to be generated from nationally-produced coal, and to benefit from a 1 Pta/kWh subsidy.

In addition to these policies that directly affect electricity fuel inputs, a number of others indirectly affect the potential market for renewable energy: imposing a carbon tax will favour renewable electricity generation, if fossil fuel inputs to electricity generation are not exempted. However, an energy tax which does not distinguish among the different carbon intensities of different fuels would not provide the same incentive for renewable energies.

Some IEA countries, although initiating policies explicitly to promote renewables, have an electricity pricing structure that actually discriminates *against* renewables in niche markets where they are most competitive. This occurs when a state electricity company is obliged to serve remote and/or island communities with power at a price equal to that on the mainland. This is the case in France – which explicitly recognises that this is a problem in renewable electricity promotion. A barrier can also take the form of a regulated or limited price increase, such as in Greece. In this situation the electricity provider is not allowed to reflect its supply costs in its tariffs, although the cost of supplying electricity to consumers not connected to the mainland electricity grid is higher[22]. The price charged for electricity in islands or other remote locations may therefore be lower than its cost, with the supplier's loss being recouped via cross-subsidisation from electricity users in more densely populated areas. Both France and Greece have islands on which electricity demand is significant and where solar and wind resources are good. However, the use of wind and solar electricity is relatively recent and still small-scale and the majority of electricity generation diesel-based. Any change in pricing structure to bring the price of island or off-grid electricity more in line with its cost would help to promote renewable electricity (although such a pricing revision may not be undertaken because of public service provision requirements on the electricity utility).

---

22 However in the case of the French West Indies, the electricity producer subsidises solar hot water heaters to reduce electricity demand (and therefore the amount of electricity sold at a loss).

Another case of significant off-grid electricity demand is in the Australian outback, where there are good solar resources. However, PV systems in Australia are promoted only via information campaigns, and limited low-interest loans, and their adoption has been relatively small.

Policies that have recently been used successfully to increase renewable electricity production have included:

- creating a protected market for certain amounts of renewable electricity as in the UK's NFFO;

- providing a guaranteed market *and premium prices* for any renewable electricity generated as in Germany;

- mandating electricity purchase for a certain time at fixed "avoided cost" levels (e.g. the PURPA in the US, where certain "qualifying facilities" obtained a 10 year guaranteed electricity market at utility avoided cost rates – although this would no longer work as a promotional policy for renewables as avoided cost levels have fallen sharply);

- providing capital subsidies for renewable electricity systems and guaranteed markets at favourable prices for renewable electricity as in Spain;

- allocating tax rebates or exemptions for renewable electricity systems, investments or equipment as in Finland, and

- agreeing siting plans for wind turbines with local authorities as in Denmark – although this is combined with other incentives including output subsidies.

None of these policies is completely in line with unfettered competition. "Green pricing", which is a completely market-based policy measure, has taken off in some IEA countries, but its overall impact on renewable electricity production has been limited.

COMPARATIVE ASSESSMENT OF DIFFERENT IEA COUNTRIES' RENEWABLE ENERGY POLICIES

Annex B of this report details the policies in place in each IEA country that are used to promote renewable energy. Some policy types such as competitive bidding procedures and "green pricing" are rapidly becoming more common. Increased information availability and dissemination has resulted in countries learning from each others' renewable energy policy successes. This can be illustrated by France and Ireland, both of which initiated competitive bidding procedures for renewable electricity projects of a type similar to that found in the UK's *Non-Fossil Fuel Obligation,* after the NFFO had been in operation for a number of years. Moreover, developers of the French and Irish policy learnt from early NFFO orders, and instigated policies that guarantee prices for up to fifteen years (rather than six or eight, as in early rounds of the NFFO)[23].

Although "green pricing" schemes have been available in parts of the US for a number of years, it was the Netherlands which first introduced the concept in Europe, in 1996. By early 1998, distributors in Germany, Sweden, Denmark, Switzerland and the UK had followed suit. (This policy type is not available nationwide in any country and is the policy type that is the least homogenous among or even within countries: every electricity distributor seems to have different rules by which green pricing operates).

One area of renewable energy policy that is surprisingly rare is the explicit evaluation of the successes of renewable energy policies. Switzerland and Spain carry out regular evaluations of progress towards their target, and Ireland carried out a thorough evaluation of its renewable energy policy after the first round of its Alternative Energy Requirement. However, systematic evaluation of policies is the exception rather than the rule.

---

23 When the period over which there is a guaranteed electricity market increases, developers can bid lower electricity prices, as their amortisation period is longer.

9

CONCLUSIONS

Renewable energy is a carbon-free or carbon-neutral energy source, the increased use of which can be part of the package of policies and measures by which IEA countries work to meet their emissions commitments as set out in the Kyoto Protocol. Increased renewable energy supply can also help governments achieve other policy goals, including energy security and diversity.

"Renewable energy" encompasses a wide diversity of fuel types and fuel uses. The applicability of each of these, and the cost of different renewable energy technologies vary greatly. Some renewable energy sources, such as biomass, are widely used and widely available. Others, such as geothermal, are widely used where they are available. Some renewable energies such as solar and wind are not yet used on a large scale, although they have a huge untapped potential.

The current costs of harnessing renewable energy, which are generally higher than those of other energy sources, means that government intervention will continue to be necessary if governments are to give increased importance to renewable energy. One of the most promising areas for increases in renewable energy supply is renewable electricity (especially from wind, biomass and wastes), and many IEA governments emphasise renewable electricity in their renewable energy promotional packages.

Although the importance of renewable electricity use has been growing, the liberalisation of electricity markets in many IEA countries creates many uncertainties for renewable electricity. Liberalisation has already affected the policies by which renewable electricity is being promoted in some IEA countries. None of the successful policies that have recently been used to promote renewable electricity has been compatible with completely free electricity markets. Liberalisation may, indeed, have a negative impact on renewable energy growth. For example, in Italy, pre-liberalisation promotional policies for renewables were scrapped on competition grounds.

If the overall effect of market liberalisation on renewable electricity is negative, many IEA countries will not reach their stated aim of increasing renewable electricity generation. This could also result in emissions from the power sector being higher than expected. And it may make the attainment of certain environmental goals, such as those agreed to at Kyoto, more difficult.

Some market imperfections may help a government achieve policy aims, including social objectives. However, market imperfections may be in contradiction with other stated objectives, such as increased renewable energy use and reduced emissions of greenhouse gases. Examples of market imperfections that inhibit renewable electricity use include the provision of subsidies for other electricity inputs or political decisions by national utilities to develop a certain fuel mix, without due consideration to the costs such a decision entails.

There are limits to the penetration of intermittent renewable electricity into electricity systems. Some of the reasons for this are technical, such as the need to balance changes in demand with variations in supply: others are economic, such as the costs of grid-connection, grid-strengthening and capacity backup. There is no consensus about the maximum level of renewable electricity's importance, with different groups and different countries estimating a limit that varies between 5% and 25% of a system's total electricity supply. Denmark was approaching the lower limit in 1996 and the growth in importance of wind electricity in Denmark is likely to continue, but Denmark operates in the broader context of Nordic electricity trade. The reliance of all IEA countries other than Denmark on intermittent electricity sources is significantly under 5% of electricity supply, so this issue is not likely to be a significant barrier for wind and solar electricity in the medium term, especially as international interconnections become more important and allow the smoothing of intermittent power production to a certain extent.

Declining renewable energy costs, improvements in renewable energy technology, and governments' interest in promoting renewable energy because of their positive environmental effects, should result in the continued growth of renewable energy use and importance. This is likely to be particularly marked for those renewable energy sources which are most cost-competitive now, notably wind electricity,

---

Policy conclusions

Countries should:

☐ **work towards reflecting the cost of renewable energy's environmental benefits in energy prices.** Although this aim has been agreed to by IEA governments in their "Shared Goals", little progress has been made to date in most IEA countries despite the negative externalities associated with other forms of energy. This is partly because the means to achieving this goal (e.g. via carbon taxes) are politically difficult and partly because of the sheer complexity and uncertainty of the calculations.

☐ **support the learning curve for renewable energies**, through R, D & D or through incentives for increased renewable energy deployment. Competitive bidding procedures in a ring-fenced market are a good example of effective support for increased deployment: this encourages competition to reduce the cost of harnessing renewable electricity, while at the same time limiting government expenditure on renewable energy support.

☐ **empower end-user choice**, which would allow consumers to buy "green" electricity.

whose cost is rapidly declining. The use of renewable energy has already helped to reduce growth in greenhouse gas emissions from the energy sector, and many IEA governments see an increased use of renewable energy as both feasible and desirable. While consumers can help by increasing their demand for "green electricity", it is largely up to governments (at least in the short term) to provide a positive environment in which renewable energy can become increasingly used, and increasingly competitive.

10

COUNTRY CHAPTERS

AUSTRALIA

OVERVIEW OF RENEWABLE ENERGY POLICY

Enhancing energy efficiency and renewable energy use is one of the main planks of Australian energy policy. Several policies that aim to promote renewable energy have been enacted on a Federal and state level since 1992. These policies reflect the Federal government's philosophy that the best way to allocate resources is by a fair, open and competitive market. Promotional policies therefore concentrate on addressing market barriers and failures and emphasise R&D, demonstration projects and education/information campaigns. Financial incentives to ease commercialisation and to increase the penetration of renewable electricity technologies (PV, solar thermal, wind and biomass) either by capital or output subsidies are not in place in Australia, unlike the majority of IEA countries, although limited financial incentives are available via a reduction in sales tax for renewable energy equipment, low-interest loans for solar water heaters and some other energy efficient and renewable energy applications, and exemption of alternative transport fuels from excise tax.

The three major reasons stated behind the development of renewable energy are to help mitigate greenhouse gas emissions, to contribute to domestic energy demand and to provide export opportunities for Australian business. However, Australia is rich in fossil fuels and uranium, which accounted for approximately 94% of total energy supply in 1996. Nevertheless, Australia also has a favourable climate for solar applications as well as significant potential for other renewables.

Despite this favourable climate and a high proportion of the land mass in remote areas (where the most cost-competitive niche market for renewable energy systems is) non-hydro renewable energy use in 1996 accounted for only 4.9% of TPES. Future growth of renewables in Australia is uncertain, because of the limited promotional incentives in place, the deregulation of the national electricity market and low national energy prices.

POLICIES

Renewable energy promotion is an objective of federal Australian energy policy, and state-level promotional policies have also been developed. Policies in place to encourage renewable energy promotion explicitly pursue environmental commitments as well as regional, social and rural objectives. In November 1997, the federal government announced a target of an additional 2% of electricity from renewables by 2010, as well as other programmes designed to promote increased use of renewables. Over the past few years, the federal and state governments have been pursuing micro-economic reforms of both the national energy market and the electricity industry – state governments have extensive jurisdiction over energy policy. The federal government is actively accelerating reforms of the energy industry and provided 5.6 M (A$) for this purpose as part of their climate change package. One direct impact of these reforms on renewable energy is the provision of opportunities for retailers to offer "green" electricity.

Renewable energy is cited in Australia's *National Communication*[24] as one of the main initiatives in working towards a stabilisation of GHG emissions at 1988 levels by 2000 and a further 20% reduction on this by 2005. The Communication lists the policies to promote renewables that were in place at the time of publication, i.e. the *National Greenhouse Response Strategy* and related policies (see below).

No single federal integrated national renewable energy programme covering the range of various resources is in place. However, several individual programmes have been implemented since 1992. These are generally centred around R&D or demonstration plants, information/education campaigns, and tax exemptions for renewable energy equipment and biofuels. For example the *National Greenhouse Response Strategy*, NGRS, (1992) includes sub-programmes that aim to improve the marketing of renewables, such as the *Renewable Energy Promotion Programme*, a 3M (A$) information programme which ran from 1992-1995 and encouraged the use of commercially viable renewable energy technology by publicising displays of and promoting stand-alone rural renewable energy systems. The 6M (A$) *EnergyCard* programme came into operation in May 1995 and aims to demonstrate the feasibility of innovative financing for solar water heaters and some other energy efficient and renewable energy appliances via low interest loans. The *Greenhouse 21 C: A plan of action for a sustainable future* (1994, follow-up to the NGRS) includes a *Renewable Energy Initiative* of 4.8M (A$) over 4 years to support development of a robust renewable energy technology industry and a commitment of 10M (A$) over 7 years to support "the establishment of a *Cooperative Research Centre,* with a primary objective to develop ... renewable energy."

Other federal programmes were announced in the November 1997 response to climate change. These include programmes aiming to increase the use of renewable energy such as the 21M (A$) *Renewable Energy Innovation Investment Fund* (REIIF), which will provide funds to help with the commercialisation of renewable energy technologies that are in an early development stage. Loans and grants for strategically important renewable energy initiatives will be provided by the 29.6M (A$) *Renewable Energy Technology Commercialisation* programme, and a further 10.5M (A$) has been set aside for the promotion of a few "showcase" projects.

---

24 *Climate Change: Australia's national report under the United Nations Framework Convention on Climate Change*, 1994

Tax-breaks for industry-funded R&D are available as 125% of such expenditure is tax-deductible. In 1994-95, industry R&D funds (6.92M A$) exceeded those of the government and tertiary sectors combined (5.35M A$). There are also tax incentives for biofuels: ethanol-based transport fuels are exempt from excise taxes. The Commonwealth government is also developing a white paper on sustainable development, to be produced in 1998, and will assess the adequacy of existing renewable initiatives as part of this work. Information dissemination on renewable energy information is ensured via a government-funded renewable energy Internet site.

Federal energy-related R&D activities, including renewables R&D, were carried out by the Energy Research and Development Corporation (ERDC) with additional indirect support from CSIRO (a government agency) and universities. However, severe budget constraints led the government to abolish ERDC in May 1997, as the government considered that ERDC had established a good foundation for industry to take on a greater role in funding its research. A transport biofuel programme – the *Ethanol Production Bounty Scheme*, ran from July 1994 to August 1996 and provided 3M (A$) to subsidise ethanol production; approximately 4M (A$) was committed over two years towards an R&D programme.

In addition to these federal policies, some states have initiated state-wide policies to promote renewable energy. Under the Australian constitution, states own and manage their energy resources, have different legislation and regulations governing energy supply, and have electricity utilities of varying regulatory structure. Although the separate electricity grids of the three south-eastern states are interconnected to a degree, the grids only extend over a small proportion of the country (i.e. they do not include all areas with high solar insolation). Tasmania, which is completely hydro-based, is not connected to grids on the mainland.

Current state initiatives to promote renewable energy include:

■ legislation passed in New South Wales (NSW) in December 1995 requiring electricity distributors and retailers to develop 1,3 and 5 year plans for purchasing renewable energy as part of the procedure required to obtain an (obligatory) licence;

■ a fund of up to 39M (A$) over three years to encourage the development of renewable energy, energy efficiency and cogeneration technologies in NSW;

■ the NSW government's sustainable energy development authority (SEDA) regulates green power schemes in that state: since inception of these schemes in April 1997, approximately 15000 domestic customers and 800 businesses have joined and there has been an investment of 26M (A$) in renewable energy projects associated with the schemes; and

■ an Agreement between electricity suppliers and sugar mills in Queensland to produce bagasse-based electricity (initial capacity of 49 MW possibly rising to 200 MW by 2010);

■ a solar cell initiative in New South Wales involving Pacific Power and academia that will provide 46M (A$) over 1995-1999.

On-going reform of electricity markets may encourage renewable electricity if it encourages stand-alone applications for consumers in remote locations. Increased competition in generation may also lead to construction of more modular electricity plants. There are, however, no Federal capacity quotas or purchase price guarantees for renewable electricity systems.

Some state-based support for renewable energy promotion focuses on stand-alone rural systems. The government have estimated that only around 15% of the 20,000 permanent residences and around 10% of the 60,000 holiday homes where renewable energy supply would be competitive with conventional supply actually use renewable energy systems. The promotional activities in place centre around increased information availability.

STATUS OF RESOURCE EXPLOITATION

Non-hydro renewable energy sources accounted for 4.9 Mtoe in 1996 (4.9% of Australia's total energy supply in the same year), up from 4.0 Mtoe (4.6%) in 1990. Apart from a small use of solar energy through solar water heating and PV systems, this was exclusively made up of biomass and wastes, the majority of which was used directly in homes and industry. In addition, five of Australia's eight states and territories generate electricity from hydropower, which accounted for 15.4 TWh or 8.7% of total generation in 1996. Renewable electricity generation will have to more than double in order to reach the government's target of an additional 2% of electricity generation by 2010.

Biomass

The use of biomass (mainly wood and vegetal waste, with small amounts of black liquor) has grown 22% since 1990 to reach 4.6 Mtoe in 1996. This is projected to continue growing slightly to 2000 and beyond. The largest use of biomass energy was in the residential sector (where it has oscillated at around 1.9 Mtoe since 1990), with the remainder split mainly between three industrial sectors (food and tobacco, wood production and pulp and paper) and electricity production. The fastest growth of biomass energy has been in the food industries, where use of biomass (mainly vegetal waste) has increased almost 50% between 1990 and 1996, when use stood at 1.4 Mtoe.

The Australian Government has agreed to a national target of 50% reduction of waste going into landfill by the year 2000. In 1996, landfill gas was used to generate small quantities (171 GWh) of electricity in six states/territories.

Waste

In 1996, 203 ktoe of industrial waste was incinerated in the chemical and petrochemical industries. No utilisation of municipal waste was reported.

Wind

Southern and western Australia and Tasmania have good wind resources, and in Southern Australia these resources are located relatively near the electricity grid. Non grid-connected turbines have been operating since 1991 in remote locations of these states as an alternative to oil-powered electricity. Grid-connected wind electricity was first reported at 4 GWh in 1994 (from 2.2 MW capacity in the same year) and stood at 7 GWh in 1997.

Solar

Australia's climate and the maturity of solar hot water technology has rendered such systems commercially competitive in many parts of Australia. The direct use of solar thermal energy contributed 86 ktoe to Australia's TPES in 1996. Sales of solar water heating systems have stabilised at 16-17000 per year, with approximately half of Australian production exported.[25]

Hydro

Electricity production from hydropower stood at 15.4 TWh (8.7% of total generation) in 1996, and has been at approximately this level since 1991. This varies significantly by region, with Tasmania generating all its electricity from hydropower, and the states of Victoria and New South Wales using hydro to generate 2% and 1% respectively (other states did not use hydropower at all). The largest hydro development is the 3.7 GW Snowy Mountains scheme which exports electricity to NSW and Victoria. Opportunities for further expansion are limited to some small scale projects, as most of the commercially favourable sites have been exploited and there are no real large-scale hydro potential outside Tasmania and New South Wales. Public opposition to further hydro development is also important in Tasmania. Hydro generation in 2000 is therefore expected to grow only slightly from its 1996 level to stand at approximately 17 TWh, although there may be a small rise post-2000.

---

25   National Communication: 55

Geothermal, Heat pumps

No details reported.

Table 1
Trends in renewable energy supply and use

	Unit	1990	1992	1995	1996	2000	1990-1996 (%)[1]	1996-2000 (%)[1]
Renewable TPES (excl. hydro)	**ktoe**	**4042**	**3722**	**4515**	**4903**	**5000**	**3.3%**	**0.5%**
Percentage of TPES	%	4.6	4.2	4.8	4.9	4.5		
Geothermal	ktoe	0	0	0	0	0	0	0
Solar, Wind, Wave, Tide	ktoe	81	81	84	87	90	1.1%	0.9%
Biomass and Wastes[2]	ktoe	3961	3641	4431	4816	4910	3.3%	0.5%
- Biomass	ktoe	3776	3460	4244	4614	n.a.	3.4%	n.a.
- Wastes	ktoe	185	181	187	203	n.a.	1.5%	n.a.
Renewable electricity generation (excl. hydro)	**GWh**	**600**	**2721**	**2881**	**3061**	**n.a.**	**31.2%**	**n.a.**
Percentage of total generation	%	0.4	1.7	1.7	1.7	n.a.		
Geothermal	GWh	0	0	0	0	n.a.	n.a.	n.a.
Solar, Wind, Wave, Tide	GWh	0	0	30	32	n.a.	n.a.	n.a.
Biomass and Wastes[2]	GWh	600	2721	2851	3029	n.a.	31.0%	n.a.
- Biomass	GWh	600	2721	2851	3029	n.a.	31.0%	n.a.
- Wastes	GWh	0	0	0	0	n.a.	n.a.	n.a.
Renewable TFC (excl. hydro)	**ktoe**	**3334**	**3014**	**3759**	**4079**	**4600**	**3.4%**	**3.0%**
Percentage of TFC	%	5.7	5.2	5.9	6.2	6.3		
Geothermal	ktoe	0	0	0	0	0	n.a.	n.a.
Solar, Wind, Wave, Tide	ktoe	81	81	81	84	90	0.6%	1.7%
Biomass and Wastes[2]	ktoe	3252	2932	3678	3995	4510	3.5%	3.1%
- Biomass	ktoe	3067	2751	3491	3792	n.a.	3.6%	n.a.
- Wastes	ktoe	185	181	187	203	n.a.	1.5%	n.a.
Hydro TPES	ktoe	1217	1320	1366	1323	1480	1.4%	2.8%
Hydro electricity generation	GWh	14148	15348	15885	15381	17213	1.4%	2.9%
Percent of total generation	%	9.2	9.6	9.2	8.7	8.5		

Notes:
1. Annual Growth Rate
2. Including Animal Products and Gases from Biomass

AUSTRIA

OVERVIEW OF RENEWABLE ENERGY POLICY

Hydro and non-hydro renewables provide approximately a quarter of Austria's energy needs – the fourth highest level of renewables use in the IEA. The use of biomass and wastes is particularly high, at 13.4% of Austria's energy supply in 1996. Actions towards increased use of biomass, and to a lesser extent, hydro, form the cornerstone of Austria's renewable energy policies and also play a significant role in climate and agricultural policy considerations[26].

Increased renewable energy use is encouraged at both the national and provincial level. National promotional policies emphasise capital subsidies and R&D. Since July 1997, incentives for renewable electricity generation have been provided by the *Promotion Instrument for Electricity from Renewables*. Before this, a voluntary agreement between the Ministry of Economic Affairs (MEA) and the Association of Electric Utilities on buy-back rates also provided premium buy-back rates for three years for renewable electricity systems installed before end 1996, especially those based on PV and wind. Using renewables other than to produce electricity is also encouraged, both for heat production (via federal subsidies), and solar collectors and heat pumps (via provincial subsidies). The importance of information dissemination in increasing renewable energy use is also recognised by the Austrian government.

Provincial actions to promote renewable energy use are also important. Some provinces have targets for increased renewable energy use, and provide capital subsidies for solar thermal and biomass technologies as well as heatpumps.

POLICIES

Energy security, energy efficiency and renewable energy promotion are central aspects of Austrian energy policy. Increased use of renewable energies is also the third measure listed as a response option in Austria's climate change policy. This policy includes several detailed quantified references to renewable energy potentials and reveals a relatively mature strategy for the development and promotion of renewable energy which interlinks with programmes on energy efficiency, CHP and district heating.

---

26 Austrian Energy Policy is described and analysed in detail in the IEA's forthcoming *Energy Policies of Austria, 1998 Review*.

Both the federal and provincial governments have responsibilities for aspects of energy policy, both can enact energy legislation, and both are implementing measures to promote renewables. The main forms of support are the *Promotion Instrument for Electricity from Renewables* (PIER), which provides both a capital subsidy and a guaranteed tariff for renewable electricity fed into the grid, and R&D measures. The capital investment subsidies that were in place for district heating schemes via the *District Heating Promotion Act*, including those based on biomass, were stopped at the end of 1996.

At the federal level, the Ministry most concerned with renewable energy is the MEA, although the Ministry of Agriculture and Forestry is involved in some biomass-related projects (as increased biomass use dovetails with some agricultural policy objectives) and the Ministry of Science and Research is involved in renewables-related R&D.

The PIER is used to promote renewable electricity from biomass, wind and solar electricity via a mixture of capital and output subsidies. Projects are selected for subsidy via a public competitive tender procedure, and subsidies are awarded for each technology type on a lowest capital cost basis. Thus similar renewable energies compete amongst themselves, but not between one another (i.e. while two proposed biomass schemes would compete with one another, neither would compete with a wind or solar project). The capital subsidies are capped to provide no more than a 7% rate of return for 15 years. In addition, the electricity from these systems benefits from a guaranteed market at a guaranteed buy-back-rate (based on the long term marginal costs of electric utilities) for 15 years. The federal government incentive structure has also been altered so that any incentives received from the provinces are now taken into account when calculating the level of federal subsidies.

Federal subsidies for biomass-based district heating were 192.4 M ATS[27] in 1996 (compared to 142.8 M ATS in 1995) and were disbursed under two programmes. One, administered by the Federal Ministry for Agriculture and Forestry that also includes incentives for biofuel production, disbursed 134 M ATS in 1995, and 189 M ATS in 1996, see below. This was augmented under the *District Heating Promotion Act* by 8.9 M ATS in 1995 and 2.6 M ATS in 1996. Capital subsidies of 8% are available for the construction, restoration or enlargement of small (0.5-10 MW) hydro systems, and subsidies of up to 25% are available for micro (<0.5 MW) systems that are both remote and environmentally friendly. In addition to federal promotional measures, provincial authorities may also provide subsidies or favourable loans for biomass heating plants. These incentives have been most favourable for agricultural operators of such systems, which has had the additional benefit of helping farmers increase their income in some of the less touristy provinces of Austria.

---

27   In 1997, 1 US $ = 12.2 ATS

Agricultural subsidies of 190 M ATS in 1996 administered by the Ministry of Agriculture and Forestry were used for the following renewable energy purposes:

- converting fossil fuel-based heating installations to biomass-based heating installations;

- replacing obsolete heating installations with biomass-based systems;

- biogas installations;

- installation of small hydropower installations (up to 200 kW);

- small-scale installations for the production, transmission and distribution of biomass-based district heat;

- installations for the production of liquid biofuels used on farms.

However, interest in biogas installations and the production of liquid biofuels for transport is limited due to their high production costs and the current low energy price.

The *Dwelling Improvement Act and Housing Promotion Act* provides financial support for non-hydro renewables, particularly solar, via grants or low-interest loans. Although there is no federal support at the federal level for heatpumps, financial support is provided by the provinces. For example, three provinces grant subsidies of ATS 5000 - ATS 30000 for heat pump installations in the residential sector. The provinces also support solar and biomass programmes. Typical support is in the order of 20% of total costs, and total support for solar, biomass and heatpumps was estimated at 500 M ATS in 1996 (up from 200 M ATS in 1992).

The *Solar Energy Program* ran from 1992-1995 and provided capital subsidies for PV systems and electric vehicles. The programme was financed by utilities, federal and provincial governments, and total subsidies disbursed under this programme have been estimated at 20-25 M ATS, of which 2 M ATS was for electric vehicles (200 were in operation by the end of 1995). Although only around 400 kW of PV had been installed by the end of 1994, the programme was judged successful in raising the profile of PV to the interested public.

Between 1994-1996, a voluntary agreement between the MEA and the Association of Electricity Utilities supported renewable-based grid electricity systems. The agreement provided subsidies of 100% of the delivery price for PV and wind power, and 20% of the delivery price for biomass and biogas for a period of three years. Since the programme ended in 1996, the payment of higher buy-back rates at these levels will cease at the end of 1999. Future renewable electricity tariffs and the future framework for promoting renewable electricity are being developed in the context of the EU electricity directive.

An Ordinance of the Federal Minister for Foreign Affairs in force since the 1st August 1995 guarantees minimum prices for electricity that is traded between provinces and has been produced from autoproducer CHP stations and renewable electricity plants. For plants up to 2 MW these prices range from ATS 0.421/kWh to ATS 0.9/kWh (100-125% of the prices charged by the *Verbundgesellschaft*), depending on the time of delivery. For plants over 2 MW, the rates are 100% of the prices charged by the *Verbundgesellschaft*, or ATS 0.421-0.72/kWh.

Approximately 25% of Austria's Federal energy R&D budget in 1996 was allocated specifically to renewable sources (down from 32% in 1995, but still around triple the IEA average) split 52% for biomass and 27% for solar, with the remainder spent on wind, geothermal and hydro. Biomass funding in 1996 was 50 M ATS and includes projects on the direct use of biomass (e.g. by improving wood-burning stoves), as well as the production of liquid biofuels. Solar research is directed predominantly towards heating and cooling, and photovoltaics.

Both direct and indirect information measures to support renewable energy are in place. Federal support for third party energy information and advice (consulting firms) was made available in 1993 and has resulted in estimated energy savings of 1.25 PJ per year at a cost of 5 M ATS. Support for smaller energy users was introduced in 1993.

STATUS OF RESOURCE EXPLOITATION

Non-hydro renewable energy accounted for 3.69 Mtoe or 13.6% of Austria's energy supply in 1996: one of the highest in the IEA, and more than triple IEA average. Hydropower accounted for a further 13.2% of total energy supply (and generated 66.5% of Austria's electricity in the same year). Heat production from renewable energy sources is also important, and was 4706 TJ (11% of total heat production) in 1996. By far the most important non-hydro renewable is wood, almost all of which is used directly in the residential and agricultural sectors. Other biomass is also used in industry and for electricity and heat generation. Wastes are also used for electricity and heat generation, and limited quantities of geothermal, solar and wind energy are also harnessed.

Biomass

Biomass provides approximately 12% of TPES: significantly higher than the IEA average. In 1996, 3.6 Mtoe of biomass (wood, wood waste, biogas and diester) were used for energy purposes. The vast majority is reported as direct use under TFC, with the remainder used for generation of electricity and heat.

Firewood accounted for the majority of biomass energy use. The use of wood chips is also increasing, and 2528 heating systems using wood chips were installed in 1996, bringing the total to 20,328 (with a total capacity of 2 GW). In addition, Austria produced 12.9 kt of rapeseed methyl ester in 1996.

Wastes

Austria used 197 ktoe of municipal and other solid wastes in 1996. These were used for electricity and heat production, and were also used directly in the buildings sector and industry.

Wind

Wind electricity generation is small in Austria. Prototype turbines were constructed in 1994, and by 1996 total wind capacity was 11.7 MW. Generation was 4.6 GWh. Capacity is growing rapidly, and stood at 13.2 MW in mid-1997 with a further 55 MW in the planning stages.

Hydro

Electricity production in Austria has historically been dominated by hydropower, which provided 66.5% of Austria's electricity in 1996. Utilities accounted for the majority of this, although there is also some autoproduction. The number of small hydro systems has been growing, and made up 8.8% of total hydro capacity. Hydropower has climbed steadily between 1990 and 1995 (although low precipitation caused a drop in 1996). Hydropower is likely to retain its dominance in Austrian electricity supply as there is room for capacity expansion: only an estimated 70% of economically exploitable hydropower sites are currently being used.

Solar

Austria uses solar collectors to heat water, and a small amount of PV systems to generate electricity. The area covered by installed solar collectors reached 1.5 M (m²) in 1996 (compared to 0.6 M (m²) in 1991) approximately two-thirds for space and water heating and one third for swimming pool heating. The annual heat output in 1996 was estimated at 1764 TJ. The number of photovoltaic systems has been growing rapidly and was 4200 in 1996, with the majority (3700) not grid connected. However, the total contribution of solar power to Austria's energy balance in tiny, and unlikely to grow significantly in the near future.

Geothermal

Geothermal energy is used in district heating and for heat production in the Southern Styrian/Burgenland basin and Upper Austrian molasse zone. Installed capacity in 1996 was 21.1 MW from six installations.

Heat pumps

The use of heat pumps in Austria is growing, and there were approximately 130,000 heat pumps installed by the end of 1996. This corresponds to 632 MW and a heat production estimated at 5595 TJ. Heat pumps are used mainly to heat water but also for space heating and heat recovery. A critical problem faced by Austria is the number of manufacturers and models (more than 2000) which has not facilitated economies of scale and cost reductions, even though large amounts of public money were invested to promote heat pumps.

Table 1
Trends in renewable energy supply and use[1]

	Unit	1990	1992	1995	1996	2000	1990-1996 (%)[2]	1996-2000 (%)[2]
Renewable TPES (excl. hydro)	**ktoe**	**2752**	**2541**	**2717**	**3686**	**3660**	**5.0%**	**-0.2%**
Percentage of TPES	%	10.7	9.9	10.3	13.6	12.6		
Geothermal	ktoe	0	0	0	4	0	n.a.	n.a.
Solar, Wind, Wave, Tide	ktoe	0	0	0	43	50	n.a.	4.1%
Biomass and Wastes[3]	ktoe	2752	2541	2717	3639	3610	4.8%	-0.2%
- Biomass	ktoe	2729	2518	2642	3442	n.a.	3.9%	n.a.
- Wastes	ktoe	23	23	75	197	n.a.	42.7%	n.a.
Renewable electricity generation (excl. hydro)	**GWh**	**1116**	**1292**	**1957**	**1763**	**2700**	**7.9%**	**11.2%**
Percentage of total generation	%	2.3	2.5	3.5	3.3	4.8		
Geothermal	GWh	0	0	0	0	0	n.a.	n.a.
Solar, Wind, Wave, Tide	GWh	0	0	0	5	0	n.a.	n.a.
Biomass and Wastes[3]	GWh	1116	1292	1957	1763	2700	7.9%	11.2%
- Biomass	GWh	1116	1292	1870	1682	n.a.	7.1%	n.a.
- Wastes	GWh	0	0	87	81	n.a.	n.a.	n.a.
Renewable TFC (excl. hydro)	**ktoe**	**2543**	**2298**	**2298**	**2440**	**2810**	**-0.7%**	**3.6%**
Percentage of TFC	%	12.1	10.6	10.6	11.0	11.2		
Geothermal	ktoe	0	0	0	4	0	n.a.	n.a.
Solar, Wind, Wave, Tide	ktoe	0	0	0	42	50	n.a.	4.3%
Biomass and Wastes[3]	ktoe	2543	2298	2298	2395	2760	-1.0%	3.6%
- Biomass	ktoe	2520	2275	2275	2328	n.a.	-1.3%	n.a.
- Wastes	ktoe	23	23	23	67	n.a.	19.1%	n.a.
Hydro TPES	ktoe	2708	2995	3188	3909	3390	6.3%	-3.5%
Hydro electricity generation	GWh	31494	34831	37065	35580	39400	2.1%	2.6%
Percent of total generation	%	63.7	68.5	67.2	66.5	70.6		

Notes:

1. Renewables data formally submitted on an annual basis by country's administration to the IEA have in some instances been supplemented with data from national publications or other sources.

2. Annual Growth Rate

3. Including Animal Products and Gases from Biomass

BELGIUM

OVERVIEW OF RENEWABLE ENERGY POLICY

The percentage of total energy supply met by non-hydro renewable energies in Belgium was 1% in 1996, compared to an IEA average of 3.9% in the same year. The backbone of Belgian activity in the field of renewables is on R&D and demonstration. There are also measures promoting renewables, such as economic and fiscal incentives via capital and output subsidies or tax exemptions, and information campaigns are also used. The national Government perceives three main barriers which inhibit increased penetration of renewable energy:

■ Belgium's limited renewable energy resources;

■ the existence of large, centralised energy production systems or networks with good grid-connections, which enables consumers to benefit from economies of scale;

■ the low relative price of conventional energies.

The promotion of renewable energy is mainly undertaken at a regional, rather than national level, with the exception of tariff-setting for electricity buy-back rates. These rates, however, have been too low for renewable energy autoproducers to increase significantly their output (except where other constraints apply such as space constraints on landfilling exist for biomass and wastes). The regions have therefore chosen to emphasise support for R&D. However, financial support is also available for demonstration and promotion of technologies reaching technological maturity and for technologies close to being economically competitive or technologies already on the market.

POLICIES

Belgian energy policy emphasises R, D & D and promotional activities in order to meet the following objectives:

■ energy security and diversity;

■ environmental protection, namely reduction of CO_2 emissions;

■ strengthening the technical capabilities of industrial exporting firms; and

■ an active involvement in EU and other international renewable energy programmes.

Belgian energy policy emphasises the national and efficient use of energy to counter high energy import dependency. Renewable energy plays only a small role in the Belgian energy scene due to its relatively small potential. Promotional activities for renewable energy are limited at a national level to setting electricity buy-back rates and to federal R&D activities, as the progressive regionalisation of Belgian energy policy has transferred responsibility and promotion of renewables to Belgium's three regional administrations (Flanders, Wallonia and Brussels).

Regional renewable energy R, D & D and promotion programmes such as economic or fiscal incentives, regulation, information and training are therefore in place, although the renewable energies targeted vary between regions. Different groups in Belgium also participate actively in European programmes that have a strong renewable energy component, such as JOULE, THERMIE and ALTENER.

Flanders has a renewable energy target of doubling renewable energy use over the period 1996-2000, and achieving a use of 5% of energy consumption from renewable sources by the year 2020. Wallonia's 1995 *Environment Plan for Sustainable Development,* which was developed as a plan to reduce CO_2 emissions, includes an aim to increase renewable energy use to 3% of energy consumption by 2000 and 5% by 2010.

Renewable energy promotion programmes are the basis of one of 14 measures cited in the Belgian *National Plan to reduce CO_2 Emissions*. The plan states that each authority (federal or regional) will seek to promote renewables within the current system of grants, subsidies or agreements. For the federal government, this has meant granting more favourable conditions for autoproduced renewable electricity (see below), although current incentives are still relatively low. The regional governments continue to rely on financial incentives, mainly grants for R&D projects. The Plan explicitly quantifies renewables' contribution to CO_2 mitigation in the industrial sector at 0.2 Mt of CO_2 (or 2.6% of total emission reductions envisaged in the plan). Potential export markets are an important driver for the country's renewable manufacturing capability in biomass and wastes, wind, hydro and PV. However, the federal government believes that stimulation of domestic renewable electricity equipment sales and generation is prevented by institutional resistance in the electricity supply industry and a specific lack of measures to overcome this.

While the realistic maximum penetration of renewable energy in Belgium is low (e.g. 5% of total energy by 2010-2020), even this is not likely to be achieved unless further financial, regulatory and information measures are adopted. One way of increasing investor interest in renewable energy is to offer more generous renewable electricity payments for a limited period, and the Belgian Control Committee for Electricity and Gas (CCEG) increased the financial incentives available for renewable electricity in 1995. There is now a payment of 1 FB/kWh produced[28] (or, on request from the producer, in the form of a capital subsidy to

---

28   On average in 1996, 1 US$ = 30.98 FB

help cover the cost of grid-connection). This "VIREG" subsidy is available for the first ten years of electricity production for renewable energy projects on line before the end of 1998. This measure will cost an estimated FB 30m per year on the basis of present production levels.

The 1995-2005 *Electric Equipment Plan* was approved by the government in January 1996 and includes measures aiming to increase the production of electricity from renewable energy resources (hydroelectricity, solar, wind, biomass and biogas). This is projected to increase renewable electricity capacity by 35 MW by 2005. However, a guaranteed market for renewable electricity does not exist in Belgium. The Belgian CO_2 reduction programme cites strengthening of *Electric Equipment Plan* as part of the climate change response strategy and the government has asked the CCEG to continue its study of tariffs that would be favourable to increased use of renewable electricity.

Renewable energy technologies are eligible for a range of existing financial support instruments for 'new technologies'. Up to 100% of research costs and 50% of demonstration costs are to be made available for renewable projects, and investments in non-R&D systems also benefit from grants and tax allowances. More detailed information is shown in Tables 1 (Flemish region) and 2 (Walloon region).

Although both the Flemish and Wallonian Governments concentrate their R&D activities on energy conservation technologies, both include some renewable technologies. Total government expenditure on renewable energy was $3.07m in 1996, down from $4m in 1995. Over half of the 1996 total ($1.57m) was directed towards solar energy – both photovoltaics and solar heating and cooling. R&D expenditure on biomass dropped sharply to $0.77m in 1996 compared to its 1995 value of $2.12m.

Further regional support for renewables includes use of information programmes to raise awareness. In 1994, the Walloon region issued a guide to promote the development of renewable energy sources, and information on renewable energy is available in the Walloon energy information centres. Flanders are also planning to set up an information centre on renewable energy. The measures to promote renewable energy that are summarised in Table 2 are those for which the Wallonia region Direction Générale des Technolgies, de la Recherche et de l'Energie (DGTRE) is responsible.

Biofuels for transport are promoted via a pilot programme in public transport vehicles in the Wallonia region and in Brussels. In 1995, the Flemish government cofinanced via the EU project ALTENER a large-scale demonstration and test programme for biofuels in cars (total funding 8.6 M BF).

Table 1

Renewable energy support measures in the Flanders Region

Name	Description	Budget
Investment grants (ecology support)	15% investment support for renewables	unlimited (part of general investment support budgets)
Demonstration projects	35% investment support for renewables	17.5 M BF
VLIET programme (1992-1995)	* grants for research projects (50-100%) * development projects (50% refundable loan, converted to 25% grant in case of success, 40% grant in case of failure)	132.5 M BF (1992-1995)
VLIETbis (1997-)	as above	67.5 M BF
Govnt. allocation for: – VITO energy research department	Flemish Institute for Technological Research	± 250 M BF (energy in general, part of which renewables)
– IMEC PV department	Research Institute	± 60 M BF
– dissemination activities (1998)	Dissemination activities, market stimulation	2.5 M BF
– firms involved in renewables	Subsidies of up to 0.1 M BF	0.5 M BF
PV subsidies scheme	50% investment support to stimulate market penetration	20 M BF
Output subsidy (VIREG)	Wind, solar and biomass electricity is subsidised 1 BF/kWh. Thermal solar units also subsidised.	249 M BF (energy in general, including renewables)
Tax allowances	14% of renewable investments can be deducted from company profits	not calculable
Research and development projects	A range of projects executed by research organisations, universities and agencies	± 40 M BF

Table 2
Renewable energy promotion in Wallonia

Measures	Description	Beneficiaries
Decree of the Walloon Region in the field of RD&D of 5 July 1990	Financial support for RD&D projects: – Subsidies for basic industrial research: up to 50% of the Eligible Cost (E.C.) (general provision); up to 80% of the E.C. for SME's; and up to 100% of the E.C. for universities and research centres – Subsidies for the preparation of RD&D projects for accompanying measures: up to 80% of the E.C. for SME's – Recoverable advances for applied research, development and demonstration projects: up to 50% (general provision); up to 80% of the E.C. for SME's or for projects meeting specific conditions	Firms (including SME's) groups of firms, research centres, universities Small and Medium Firms (SMEs) Firms including SME's
Tax Allowances (fiscal incentives)	13.5% (1997) of the allowable costs for renewable energy investments can be deducted from the profits of enterprises	Firms
Royal Decree 10.2.83 "ECHOP"	Subsidies (20%) of the E.C. for renewable energy investments	Schools and hospitals
"Arrêté" of Dec. 19, 1984	Subsidies covering specific expenses incurred by industrial enterprises: expertises (75% of the E.C.), technical agreement procedures (60%), studies (75%)	Industrial firms, groups or federations or organisations of industrial firms
Decrees of June 25, 1992 and "arrêtés" of 16 September 1993	Specific measures (e.g. awards) for investments in the field of renewables (15%)	Firms
Information centres	Publications disseminated in information centres	General Public
The S.I.P.S. (Scientific Institute for Public Services) programme	RD&D activities in the field of biomass and waste valorisation	

STATUS OF RESOURCE EXPLOITATION

Non-hydro renewable energy supplied 565 ktoe or 1% of Belgium's total energy supply in 1996, compared to an IEA average of 3.9%. Renewable energy use is almost exclusively made up of biomass and wastes, with wind, solar and geothermal use almost negligible and only growing slowly. In addition hydro generation contributes approximately 0.5% to total electricity generation (239 GWh, compared to 338 GWh in 1995).

While there is potential for expanded renewable energy use in Belgium, it is extremely limited due to the lack of large hydro resources, by limited land available for biomass plantations, and by a small coastline (where wind resources are higher). Indeed, one study gave an optimistic potential of renewable energy as 2.7-3.8% of TPES by 2010. Even this low level, however, is unlikely to be achieved unless renewable energy promotional policies are strengthened.

Biomass

Biomass accounts for just under half of Belgium's renewable energy use, and has grown slightly at a national level since 1990, contributing 231 ktoe in 1996. Most biomass is used directly for heat (mainly in the residential sector in Wallonia). Solid biomass (and black liquor in the pulp and paper factory in Harnoncourt) is also used for electricity and heat production. A small amount of biogas from sewage sludge and industrial digesters is also used to produce electricity and heat. This should increase, as a plant to extract and use landfill gas is currently under construction. Development of biomass energy is high priority in Wallonia, and electricity generation from biomass has been the fastest growing renewable energy use in Belgium over the last few years. Transport biofuels have not been produced since 1995.

Waste

Municipal and industrial wastes together accounted for an estimated electricity production of 960 GWh in 1996, compared to 617 GWh in 1990. They were also used to generate a small amount of heat. Municipal waste used to be much more widely used than industrial waste, but use of the latter has been growing rapidly and is now almost at the same level. However, the cost of generating electricity from waste has recently increased with the mandatory requirement for waste gas purification.

Wind

Wind contributes a tiny amount to electricity generation: 7 GWh in 1996 (compared to 9 GWh in 1994), mainly from Flanders. However, wind power will increase slightly (at least in Flanders) in future, with the installation of a 400 kW turbine in Hasselt and an additional 600 kW turbine in Zeebrugge. Given the topography, wind resources are limited, and significant expansion is unlikely in the near-term.

Solar

Use of solar energy has contributed just under 1 ktoe to TPES since 1990. National production of PV electricity was 58 MWh in 1996. Flanders is emphasising the development of PV, whereas Wallonia is concentrating more on solar heating and cooling and passive solar applications.

Hydro

Belgium is one of the few IEA countries whose hydrocapacity is made up mainly of small hydro plants (< 10 MW). Hydropower output was 338 GWh in 1995 but declined to 239 GWh in 1996 due to lower precipitation in that year. The 1996 capacity of 96 MW represents around two-thirds of the economically exploitable potential, with any development of further capacity limited by environmental constraints.

Geothermal

A small amount of geothermal heat (1.4 ktoe) was reported for the Wallonia region in 1996. Belgium's use of its low enthalpy resources have increased over the 1990s, and could be developed further. However, no official forecasts are available, and development of geothermal energy is low priority.

Table 3
Trends in renewable energy supply and use

	Unit	1990	1992	1995	1996	2000	1990-1996 (%)[1]	1996-2000 (%)[1]
Renewable TPES (excl. hydro)	**ktoe**	**445**	**520**	**553**	**565**	**151**	**4.1%**	**-28%**
Percentage of TPES	%	0.9	1.0	1.1	1.0	0.3		
Geothermal	ktoe	1	1	1	1	0	5.7%	n.a.
Solar, Wind, Wave, Tide	ktoe	2	2	2	2	1	0.4%	-10.5%
Biomass and Wastes	ktoe	443	517	550	562	150	4.1%	-28.1%
- Biomass	ktoe	210	228	231	231	n.a.	1.6%	n.a.
- Wastes	ktoe	232	289	319	331	n.a.	6.1%	n.a.
Renewable electricity generation (excl. Hydro)	**GWh**	**661**	**930**	**1073**	**1107**	**770**	**9.0%**	**-8.7%**
Percentage of total generation	%	0.9	1.3	1.5	1.5	1.0		
Geothermal	GWh	0	0	0	0	0	n.a.	n.a.
Solar, Wind, Wave, Tide	GWh	8	8	8	7	10	-2.2%	9.3%
Biomass and Wastes	GWh	653	922	1065	1100	760	9.1%	-8.8%
- Biomass	GWh	36	126	138	140	107	25.4%	-6.5%
- Wastes	GWh	617	796	927	960	653	7.6%	-9.2%
Renewable TFC (excl. hydro)	**ktoe**	**196**	**188**	**187**	**187**	**n.a.**	**-0.8%**	**n.a.**
Percentage of TFC	%	0.6	0.5	0.5	0.5	n.a.		
Geothermal	ktoe	0	0	0	0	n.a.	n.a.	n.a.
Solar, Wind, Wave, Tide	ktoe	0	0	0	0	n.a.	n.a.	n.a.
Biomass and Wastes	ktoe	196	188	187	187	n.a.	-0.8%	n.a.
- Biomass	ktoe	196	188	187	187	n.a.	-0.8%	n.a.
- Wastes	ktoe	0	0	0	0	n.a.	n.a.	n.a.
Hydro TPES	ktoe	23	29	29	21	30	-1.8%	9.9%
Hydro electricity generation	GWh	266	341	338	239	344	-1.8%	9.5%
Percent of total generation	%	0.4	0.5	0.5	0.3	0.5		

Notes:

1. Annual Growth Rate

2. Including Animal Products and Gases from Biomass

CANADA

OVERVIEW OF RENEWABLE ENERGY POLICY

Canadian support for renewables was initiated during the oil supply crises of the 1970s and early 1980s, as part of federal efforts to reduce oil import dependance. Support for new renewables continues to evolve against a long-term background of abundant supplies of cheap hydro electricity and fossil fuels and more recent policy developments such as electricity liberalisation and climate change commitments. Two decades of federal R&D efforts combined with measures to enhance private sector renewable energy markets in the provinces have resulted in some limited increases in the penetration of small hydro, wind, solar and biomass applications.

In 1996, non-hydro renewables supplied 3.8% of Canadian energy (8.9 Mtoe). Approximately a further 13% of energy supply is derived from hydroelectric power. Forest biomass accounts for the majority of the non-hydro renewables, with the contribution from other renewable energy sources (e.g. wind, solar, tidal) very small. Use of ethanol for transport fuel is increasing, as is the development of lower cost ethanol supplies.

The outlook for grid-based non-hydro renewables in Canada depends in particular on conditions for independent renewable electricity production in a liberalised electricity sector. Renewables will be particularly sensitive to the emerging structure of inter-provincial and the wider North American electricity market. Pressure to open up the electricity system to independent power producers and the imperative of national greenhouse gas emission reduction targets is slowly encouraging a greater focus on non-hydro renewables. Since electricity generation falls primarily within the jurisdiction of the provinces, their agreement and cooperation will be an important factor and prerequisite to designing a successful strategy.

A broad new federal strategy to promote renewables by enhancing technology, fostering market interest and providing market access for both grid and off-grid applications has been in place since 1996. The government believes that renewable energy technologies will play an increasingly important role in Canada's energy mix and will make a growing contribution to reduced greenhouse gas emissions.

Canada is the second largest country in the world with a plentiful supply of both renewable and non-renewable energy resources. Energy is important in the Canadian economy: Canada has a relatively high per capita energy demand due to a cold climate, large country and the presence of energy-intensive industries. In addition, exports of fossil fuels and electricity account for around 8% of GDP. A large proportion of electricity exports are from hydropower, which also provides approximately 64% of domestic electricity generation. Non-hydro renewables (mainly biomass use in the pulp and paper industry) account for over 4% of total energy supply. Greater use of renewable energy is seen as an important part of Canada's response to its greenhouse gas commitments that were agreed to at Kyoto in December 1997.

Renewables are promoted at both the federal and provincial levels of government. Broadly, federal involvement has been mostly in the field of R&D funding both private sector activities and performing federal R&D in fields not addressed by the private sector. Other federal measures include tax incentives and voluntary measures. Through their jurisdiction over energy production and intra-provincial distribution, provincial governments have had significant impacts over the level of deployment of renewables, especially in electricity markets.

Natural Resources Canada (NRCan) has the main responsibility for federal promotion of renewable energy, although other federal departments are also involved in renewable energy. Industry Canada and Environment Canada directly support renewable energy industry activities via their *Environmental Industry Strategy,* including international promotion. In addition, Environment Canada has R&D programmes involved in assessing wind and solar resources, and the management and energy recovery from solid waste, including landfill gas. Agriculture and Agri-Food Canada supports R&D activities related to the production of fuel ethanol from agricultural feedstock. Finally, Fisheries and Oceans Canada supports R&D activities related to the environmental impact of hydro energy.

In the late 1970s and early 1980s, because of energy supply concerns, the federal government actively promoted the development and use of renewable alternatives to conventional energy forms. Federal support peaked at about 100 M C$[29] in 1984/85, approximately half of which supported technology research, development and demonstration, and the other half as fiscal incentives. Following the collapse of oil prices in the mid-1980s, federal R&D expenditures were reduced and most non-R&D programmes were gradually phased-out. The only non-R&D programme initiated in the 1980s which is still in existence today is an accelerated capital cost allowance for certain renewable energy assets in the Income Tax Act.

29   On average in 1997 1 US $ = 1.385 C$

The majority of federal funding for technology initiatives for renewables originated from the *Program of Energy Research and Development* (PERD), a programme that conducts R&D in all areas of energy except nuclear fission. Promotion of renewable energy has formed part of the government's Green Plan (1990 – now terminated) and the 1991 *Efficiency and Alternative Energy* (EAE) programme.

In 1996, to follow up on directions outlined in the 1995 *Canada's Action Program on Climate Change*, Natural Resources Canada (NRCan) introduced its *Renewable Energy Strategy*, a blueprint for cooperative action with other stakeholders to accelerate the development and, in particular, the commercialisation of renewable energy technologies. It is focused on emerging and environmentally-friendly renewable energy sources. The principal objective of the Strategy is to support the development of a more dynamic and self-sustaining renewable energy industry in Canada. Activities under the Strategy were regrouped under three headings: enhancing investment conditions, technology initiatives, and market development initiatives.

In April 1998, activities under the Strategy were expanded with the introduction of the *Renewable Energy Deployment Initiative (REDI)*, a three-year, C$12m federal initiative. Under REDI, NRCan encourages the use of reliable, cost-effective renewable energy systems for space and water heating and cooling. REDI includes broader federal support for market pull-type measures to enhance deployment and also allows for the provision of some financial incentives for increased use of renewable energy. Another new programme was announced by NRCan in May 1998, the *Renewable Energy for Remote Communities* programme, which aims to accelerate the deployment of renewable energy in these communities via improved information transfer and technical training as well as targeted project support.

To enhance investment conditions, the federal government offers several types of incentives. The capital cost allowance (CCA) Class 43.1 of the Income Tax Act provides an accelerated write-off (30% on a declining balance) for certain assets used to produce renewable energy. Qualifying systems include electricity generation (excluding medium and large-scale hydroelectricity) and the production of thermal energy for industrial purposes (excluding space heating). Since 1996, early intangible expenditures related to these qualifying systems benefit from the *Canadian Renewable and Conservation Expense* (CRCE) whose eligible expenses are 100% tax-deductible and can be financed through flow-through shares. Already available for similar expenses in the oil and gas and mining sectors, flow-through shares allow a company to renounce certain types of expenditures to the purchasers of these flow-through shares. Finally, under the new REDI, businesses are eligible for financial incentives of 25% (capped at C$50,000) of the capital and installation cost of qualifying biomass and solar systems for space or water heating. A similar incentive is available for federal departments.

Technology initiatives, such as funding private sector research and development activities on a cost-shared basis, remains NRCan's main area of support towards renewables with a budget of about C$9 million a year. However, with the introduction of the Strategy and of new programs, voluntary programs have gained an increased focus. These include undertaking market assessment studies to help the industry identify key markets; performing market research to better understand consumer motivation and attitude; undertaking industry-level marketing campaigns to increase consumer awareness of existing reliable and cost-effective renewable energy systems; assisting the industry with infrastructure development, for example by funding the development and implementation of technical training programs for systems installers; undertaking specific activities in key markets, such as the remote communities and international markets; and "leading by doing" within the federal government, particularly by developing a 'green' power initiative for federal departments.

Canada's ten provincial governments and two territorial governments play a key role in the deployment of renewable energy in Canada. While federal jurisdiction on the energy market includes jurisdiction over international and interprovincial trade and facilities, provincial governments are responsible for energy production and distribution within a province. Any regulatory initiatives to increase renewable energy production (e.g. by facilitating market access for non-utility generators, by establishing renewable power set-asides) has to be undertaken at a provincial level. During the 1960s to 1980s, provincial governments in Québec, British Columbia and Manitoba have undertaken a major push to develop Canada's large-scale hydroelectricity potential. As a result, approximately 60% of Canada's electricity is hydro-based.

The electric power market in Canada has been characterised during the past few decades by a few producers, usually public utilities with monopoly powers over specific geographic territories. Most provinces allow for self-generation of power (usually at avoided cost rates – there is no discernible trend towards favourable market rates). Market access by private producers started about a decade ago with all provinces developing policies and processes to increase private production to the grid. Penetration has been slow, however, because of the limited requirements for new power. In Alberta, the 1988 *Small Power Research and Development Act* provided capital assistance and a guaranteed purchase price for up to 125 MW of renewable energy capacity. Among other things, it has led to the establishment in 1994 of Canada's first large-scale wind farm, the 20 MW Cowley Ridge wind farm in southwest Alberta, which accounts for most of the wind energy capacity currently in place in Canada. In Québec, the development of small-hydro sites by private developers has been encouraged by the provincial government. Also, the provincially-owned electric utility, Hydro-Québec, will buy electricity from a 100 MW private sector wind project. The 1996 Québec energy policy announced a strengthening of these initiatives and is leading towards set-asides for private production, particularly wind energy.

Market reform in the electricity sector has started in most provinces, resulting in a series of changes impacting on the deployment of renewable energy. In Alberta, an open-access power pool has been in operation since 1997 and retail access is now being planned. British Columbia, Manitoba, Québec and New Brunswick have opened their transmission grid. And Ontario aims for full competition in the production market by 2000, with retail access to follow soon thereafter. Open access allows independent renewable electricity producers to establish projects and seek wholesale purchasers. Retail access does not currently exist in Canada but when introduced will allow product differentiation in the marketplace, for example by allowing generators and brokers to offer 'green power' to consumers. As yet, green power is not available except in a pilot purchasing project whereby some of the federal government's energy needs in Alberta is met from 'green' sources.

Other policies of provincial governments and electric utilities support the development of renewable energy sources. Many provinces and several electric utilities have programmes to provide information on renewable energy technologies and their applications, as well as programmes to support research and development activities and in some cases sales tax exemptions on new renewable energy technology.

The production and use of ethanol in low level gasoline blends is supported by the federal government and some provinces through fiscal policy. At the federal level, gasoline is subject to a 10 cent/litre excise tax; alternative transportation fuels, including the ethanol portion of ethanol-gasoline blends are not subject to this tax. This tax forgiveness is secure for the current mandate of the government. A *National Biomass Ethanol Programme* was introduced in late 1994 that makes a repayable line of credit available to registered ethanol producers, in the event that changes in the future tax treatment of fuel ethanol and other factors affect the economic viability of ethanol plants. The capitalisation of this programme is $65 million. The funding of this programme and the excise tax forgiveness are in addition to the government's *Efficiency and Alternative Energy* funding. Several provinces have reduced or removed the provincial motor fuel tax for alternative fuels, including ethanol. As an example, one of the fastest growing markets for fuel ethanol is in Ontario where the province does not apply the 14.7 cent/litre motor fuel tax to ethanol. This tax forgiveness is guaranteed for the ethanol production of new plants in Ontario through 2010.

As part of its EAE programme, the federal government funds research and development on ethanol production from cellulose containing feedstocks. These new biomass feedstocks such as straw, wood waste and municipal solid waste are attractive not only due to their low cost but also because they can use waste material for the production of an environmentally beneficial alternative transportation fuel.

STATUS OF RESOURCE EXPLOITATION

Non-hydro renewable energy supply in Canada was estimated at 8.9 Mtoe in 1996, accounting for 3.8% of the total. The majority of this was made up of biomass use in the pulp and paper industry: use of other renewable energy sources (e.g. wind, solar, tidal) is very small. Hydropower accounted for another 31 Mtoe in 1996, producing over 60% of Canada's electricity in the same year.

Total renewable energy supply including large hydro is expected to increase moderately in the foreseeable future reaching 55 Mtoe (the majority of which is large hydro) by 2020. Energy supplied by most renewable energy sources would increase although the share of renewable energy in total energy supply is projected to remain unchanged. However, this forecast is based on conservative assumptions regarding improvements to the technological and cost performance of some renewable energy technologies and does not account for the impact of NRCan's recent renewable energy strategies.

Biomass

The most important non-hydro renewable energy source in Canada is biomass. Biomass use can vary significantly between years, but in general there has been an increase of approximately 20% in the use of biomass energy since the early 1980s. Production comes mainly from the combustion of waste from forest products in the pulp and paper industries to produce steam and electricity, both for their own use and for sale to the electric grid. Biomass meets over 40% of the energy requirements of the pulp and paper industry.

In 1978, Canada began a national biomass programme with the aim of displacing oil. The target was to increase the contribution of biomass to TPES from 3% to 6% in 1990. However, the reported contribution was only 4.0% in 1990. Canada reported 8.9 Mtoe of biomass use for 1996, and expects this to increase significantly by 2010 as Canada's enormous forest resource and large agricultural production are significant potential sources of bioenergy.

In the residential sector, it is estimated that 6.6% of Canadian single family homes use wood for primary heating, and over 1 million homes use some wood for heating purposes. Other uses of biomass include electricity generation (from independent power producers). This has risen steadily since 1980 from 1300 GWh to 3714 GWh in 1996. By 2010 the figure is expected to reach 7200 GWh. Some liquid biofuels are also produced: Manitoba, Saskatchewan and Ontario together produced about 29 million litres of ethanol in 1997, for use in ethanol gasoline blends. In Canada, fuel ethanol is currently produced exclusively by fermenting grain.

Municipal Waste

No energy recovery from municipal waste was reported to the IEA for Canada. Nevertheless, NRCan estimate that approximately 140 ktoe of municipal waste was used for energy purposes.

Wind

Canada has a very large wind resource potential. Currently, wind power plants with installed capacity of about 23 MW (1996) are in operation, mostly in Alberta. Electricity output from wind was 62 GWh in 1996 and is expected to increase by a factor of 10 by 2010. Another 130 MW are being planned for installation over the next two or three years, mostly in Québec and Ontario. In addition to electrical generation, wind applications include several wind-generated water pumps in the Prairies.

Solar

Solar heating and cooling was estimated to have saved approximately 17 ktoe of other energy use in 1996. There are also more than 20,000 photovoltaic systems installed in Canada with a generation capacity of approximately 2 MW, providing a reported 3 GWh in 1996. The Canadian industry sells roughly 1 MW of new photovoltaic generation capacity per year with 50% to 60% of these sales to the export market. PV capacity is expected to rise to 5 MW by 2010, providing 7 GWh of electricity.

Active solar technology is most cost-effective for low-temperature heating applications, such as water heating, outdoor pool heating and commercial/ industrial ventilation air heating. There are an estimated 12,000 domestic solar hot-water systems and 300 commercial/industrial solar hot-water systems currently in use in Canada. A promising application of solar heating is the development of technologies to heat aquaculture water. Another potential application of solar heating is the pre-heating of air for commercial/industrial ventilation systems. Photovoltaics have increasingly become cost-effective for many stand-alone applications across Canada. These include electric power for various telecommunication systems, water pumping and purification, remote monitoring and control, remote residential, various coast-guard lighting/beacon systems and numerous consumer applications.

Hydro

Hydroelectricity (mostly large scale) accounted for over 62% of Canada's electricity production in 1996 mainly from installations in Québec, British Columbia, Newfoundland and Manitoba. In Canada, small scale hydroelectricity is normally considered to be non-utility owned facilities of 20 MW or less. Canada's total installed small hydroelectric capacity exceeds 1,600 MW. About 10% of total hydroelectricity generated was exported to the United States.

There is still a very large amount of undeveloped hydro potential in Canada. However, it is unsure whether much of this will be developed due to the remoteness of the sites, the physical difficulty of the terrain, or environmental concerns. Potential opportunities for small hydro development include rehabilitated hydro plants (operating, or abandoned), new plants constructed at existing dams, new plants constructed to serve remote communities as well as mini-hydroelectric (less than 1000 KW) and micro-hydro (less than 100 KW) stations. The technology for small hydro is well developed but there are still significant improvements being pursued in mini- and micro-hydro equipment. The economics of small hydro technology are site- and project-specific. Sites that require only repair or rehabilitation can often be economically attractive, while sites that require dam construction are generally not. The exception to the latter is in remote areas where small hydro systems are used to displace or complement diesel-generated electricity; in this case, dam construction for small hydro plants does have market potential. In recent years, small hydroelectricity has been a major contributor to the growth of non-utility generated power. Total hydroelectric output is set to increase slightly to 350 TWh by 2010 from an increase in installed capacity of 3 GW.

Tidal

Canada is one of two OECD countries (the other being France) that uses tidal energy. The low-head hydraulic demonstration plant (20 MW) was built in the Annapolis Basin of Nova Scotia in 1984. In 1996, it generated 32 GWh of electricity. There are no further plans to install additional tidal generating capacity.

Heat pumps

Approximately 35,000 ground source heat pumps are currently installed in Canada, representing about 500 MW of avoided electricity generating capacity. Over 50 schools use earth energy heat pumps. Although the capital cost of ground source heat pumps is generally higher than for other heating sources, they provide significantly reduced operating costs. In some circumstances, this can mean that the combined capital and operating costs of ground source heat pumps are less than for other energy options. This is particularly true in rural areas that do not have access to natural gas.

Table 1
Trends in renewable energy supply and use

	Unit	1990	1992	1995	1996	2000	1990-1996 (%)[1]	1996-2000 (%)[1]
Renewable TPES (excl. hydro)	ktoe	8457	8840	10448	8905	17220	0.9%	17.9%
Percentage of TPES	%	4.0	4.1	4.5	3.8	6.8		
Geothermal	ktoe	0	0	0	0	40	n.a.	n.a.
Solar, Wind, Wave, Tide	ktoe	2	3	8	8	60	22.3%	63.8%
Biomass and Wastes	ktoe	8454	8837	10440	8896	17120	0.9%	17.8%
– Biomass	ktoe	8451	8822	10440	8896	n.a.	0.9%	n.a.
– Wastes	ktoe	3	14	0	0	n.a.	n.a.	n.a.
Renewable electricity generation (excl. hydro)	GWh	2591	4128	3708	3811	11989	6.6%	33.2%
Percentage of total generation	%	0.5	0.8	0.7	0.7	2.1		
Geothermal	GWh	0	0	0	0	501	n.a.	n.a.
Solar, Wind, Wave, Tide	GWh	29	38	95	97	698	22.3%	63.8%
Biomass and Wastes	GWh	2562	4090	3613	3714	10790	6.4%	30.6%
– Biomass	GWh	2553	4048	3613	3714	n.a.	6.4%	n.a.
– Wastes	GWh	9	42	0	0	n.a.	n.a.	n.a.
Renewable TFC (excl. hydro)	ktoe	8149	8450	10122	8577	15460	0.9%	15.9%
Percentage of TFC	%	5.1	5.2	5.7	4.7	8.0		
Geothermal	ktoe	0	0	0	0	0	n.a.	n.a.
Solar, Wind, Wave, Tide	ktoe	0	0	0	0	0	n.a.	n.a.
Biomass and Wastes	ktoe	8149	8450	10122	8577	15460	0.9%	15.9%
– Biomass	ktoe	8149	8450	10122	8577	n.a.	0.9%	n.a.
– Wastes	ktoe	0	0	0	0	n.a.	n.a.	n.a.
Hydro TPES	ktoe	25527	27207	28844	30613	29950	3.1%	-0.5%
Hydro electricity generation	GWh	296827	316361	335399	352962	348310	3.1%	-0.5%
Percent of total generation	%	61.6	60.7	59.7	62.4	60.0		

Notes:

1. Annual Growth Rate

2. Including Animal Products and Gases from Biomass

DENMARK

OVERVIEW OF RENEWABLE ENERGY POLICY

Renewables have featured prominently in Denmark's 1990 and 1993 *Energy 2000* plans and continue to play an important role in the 1996 *Energy 21*, which sets out Danish energy policy[30]. The Danish government estimates that the initiatives launched in *Energy 21* will result in renewable sources of energy accounting for 12-14% of Denmark's energy consumption by 2005, compared to around 6.9% in 1996. Reducing CO_2 emissions is the principal driver of Danish energy policy, and renewables are intended to play a critically important role in replacing coal-fired, and ultimately gas-fired, power as a means of achieving this objective. Denmark has implemented various programmes to promote renewables since the mid 1970s including information dissemination, capital and output subsidies, direct regulations and guaranteed markets for renewable electricity. Domestic interest in renewable energy continues to be bolstered by Danish successes in international markets for renewable energy technology in the market for wind turbines.

In 1996, the majority of non-hydro renewable energy supply was from biomass and wastes, but with a significant quantity from wind (and a small contribution from solar energy and heat pumps). Renewable electricity (mostly wind and biomass) provided 2412 GWh or 4.5% of 1996 total electricity generation, significantly above the IEA average. For geographical reasons, Denmark has only a small hydro resource, and little potential for further development.

Denmark's success in achieving its mid-term (2005) renewable goals hinge in particular on continued development of wind power from its level of just over 1 GW at the end of 1997 to the 1.5 GW target by 2005, and on increased incineration of straw and wood chips by the utilities. The rate and nature of incremental renewable electricity capacity expansion is dependent on the integration of renewables into the Nordic electricity market in which Denmark participates. It also depends on how plans to increase the use of natural gas are managed against a background of electricity market reform and, in the longer term, the consolidation of Demark's two electricity grid systems.

POLICIES

Increased use of renewable energy sources is a core part of the Danish Government's energy policy which seeks to reduce energy consumption and

---

30 Danish Energy Policy is described and analysed in detail in the IEA's forthcoming *Energy Policies of Denmark, 1998 Review*.

emissions from the energy sector while ensuring that Danish society is furnished with a secure and environmentally benign supply of energy. Denmark aims to almost double the importance of renewable energy from its 1988 value by 2005. This will be achieved by co-ordinated energy/environment initiatives prominently featuring renewable energy beginning with the Government's original *Energy 2000 Plan* (published 1990), the *Energy 2000 Follow-Up Plan* (published 1993) and *Energy 21* (Spring 1996). The 1993 Follow-Up plan proposes a series of additional measures to be taken in the energy sector (excluding transport), to ensure the original 2000 targets will be met.

Additional measures to promote renewables in *Energy 21* include:

- ensuring that the energy supply to small communities be met by biomass-based district heating of CHP (rather than natural gas)[31];

- increasing the flexibility of regulations governing the quantities of wood chips and straw to be used in electricity generation;

- making wind turbine planning a regular feature of regional and municipal planning;

- promoting solar heating through collaboration with energy distribution companies, and

- setting up a demonstration geothermal plant.

The favourable policy climate for the development of renewables was further strengthened in 1994 by the creation of a single Ministry of Energy and Environment. This Ministry promotes renewable energy through the Danish Energy Agency (DEA) which takes operational and implementation responsibilities. A package of new initiatives for the promotion of renewables was launched by the DEA in November 1995 focusing on wind energy, but also including actions in the areas of biomass and solar energy.

In addition to its commitment under the FCCC, Denmark has set a further target to reduce CO_2 emissions from energy and transport by 20% from 1988 levels by 2005. The *Energy 2000* plan contains a series of co-ordinated measures designed with the aim of meeting the 2005 target. Various quantitative objectives including some formal targets included in the energy plans, and in Denmark's *National Communication* are listed in Table 1. (Medium term targets only are given, although the plans list some targets for 2030).

Efforts to promote renewables in Denmark continue to include regulations, economic and fiscal incentives, targets, information dissemination and R&D. Extensive evaluations of the potential for different renewable energy sources have also been carried out. Subsidies for, and R&D programmes on, renewable energy

---

31 This has, however, restricted the expansion of gas networks and of growth in gas use for other purposes.

Table 1
Medium-term Targets for Renewables Development in Denmark

Resource	Target	Current Status	Comment
Wind	1500 MW and 10% of the country's electricity by 2005	791 MW (end 1996) and >1 GW by end 1997 (slightly ahead of target)	Siting onland has been planned via local land-use plans and offshore capacity is expanding faster than expected.
Biomass	A total of 85 PJ, including mainly straw wood chips and waste per year by 2005.	59PJ (1.8Mtoe) of biomass and wastes were used in 1996.	Increased use of wood chips and straw are expected to account for the majority of the desired increase.
Solar	0.2 Mtoe active and passive solar by 2005 and 300 GWh generated by solar cells. Increase sales of solar heating units to 5000/year.	PV generation was negligible in 1996 - the majority of the 259 TJ solar energy use was heat.	Recent collaboration between the government and gas distribution companies has resulted in the installation of around 7000 combined gas and solar water heaters.

were implemented in the mid-1970s. These were strengthened in the 1980s and 1990s by targeted regulatory mandates, market stimulation measures via output subsidies and guaranteed markets for renewable electricity. Other financial incentives include reduced taxation rates for profits from the sale of renewable electricity.

Output credits (payments/kWh) were first introduced in 1983 for wind electricity and extended in 1992 to electricity from renewable CHP plants. A more favourable and extensive system of output credits for wind and other renewables including biogas and hydro, biomass and municipal waste in some circumstances, was introduced in 1992 with lower credits for small scale biomass. Different renewable technologies receive state aid in the form of output subsidies per kWh produced. These vary between 0.1-0.27 DKr[32], depending on the renewable energy source and producer (most utility producers receiving the lower rate and autoproducers the higher rate). The funds to run this subsidy scheme come from the revenues from the CO_2 tax, from which renewable energy is exempt. However, the total buy-back rate for renewable electricity is significantly higher than the subsidies, as utilities have an obligation to pay private operators 85% of the total generation and distribution costs for renewable electricity. Buy-back rates for privately generated wind electricity in 1996 averaged 0.58 DKr (including the 0.27 DKr subsidy).

Various legal measures have been used to encourage renewables. The 1976 *Electricity Supply Act* provided the first basis for direct regulation of modes of

---

32 On average in 1997, 1 $US = 6.604 DKr

generation and led to utility agreements to develop a total of 200 MW wind capacity before 1994. Measures first introduced in the 1988 *Heat Supply Act* to encourage district heating using renewables by prohibiting installation of electric heating in specified residential areas are still in force. The mandatory programme launched in 1990 to convert all local district heating plants to CHP based on natural gas or biomass should be completed by end 2000. The ban on in-field burning of straw imposed in 1990 has resulted in this being used for energy purposes. In 1992 a parliament resolution required utilities to undertake fuel substitutions, including uptake of biomass. Utilities are obliged to purchase power from wind turbine operators and providers of other types of renewable electricity at a set price. The Government has also required municipalities to elaborate plans for future wind turbine siting. In addition, from 1996, all waste must be recycled or burned in CHP plants.

The types of economic instruments employed to promote renewables have included both capital and output credits. A construction subsidy of up to 30% of approved installation costs can be given to renewable energy projects (although this was phased out in 1989 for wind). Phase outs of capital support for technologies will occur where appropriate improvements in market conditions are foreseen (e.g. active solar). Subsidies of 15-30% are available for capital investment in domestic solar hot water and space heating systems. A three year programme (1994-1996) had a total fund of 30 M DKr to subsidise the replacement of obsolete wind turbines with modern units up to 15% of cost or 200, 000 DKr, whichever is the lowest. Subsidies of up to 50% for CHP plants using biomass can be obtained from a pool of 25 M DKr/year up to 2000.

Two Danish electricity companies have recently introduced "green pricing" schemes. The scheme at Himmerlands Elforsyning was imitated in March 1998, and charges a premium of 0.05 DKr/kWh for "green" electricity. The funds received from this scheme will be used to invest in renewable electricity projects.

Because of increasing difficulty in finding sites for wind turbines onland, the government has taken several initiatives to enlarge the potential of suitable areas for wind power including: requiring municipalities to submit proposals for wind turbine capacity; a study of increasing efficient use of existing sites; and an assessment of economic and technical aspects of developing marine sites; improved access for private investors to own and operate wind turbines, and the establishment of an information strategy for wind energy. In September 1997 the Government and utilities entered into an agreement about the installment of the first batch of off-shore wind turbines with a capacity of 750 MW.

The two Danish utilities, ELSAM and ELKRAFT, implemented a bioenergy development programme in 1992. The 1993 *Biomass Agreement* requires the utilities to use 1.2 mt of straw and 0.2 mt of wood chips annually by 2000 (about 0.4 Mtoe), either in biomass-only power stations or co-fired with coal, although this approach was relaxed slightly in *Energy 21*, which allows greater flexibility in how the total biomass target is met. The *Biomass Agreement* was amended in July 1997 leaving the utilities greater choice in the type of biomass procured. The requirements are now at least 1m tons straw, 0.2mt wood chips and 0.2 mt of either straw, wood chips or willow: this should minimise the need for expensive imports of straw from neighbouring countries. Total biomass incineration should be 19.5 PJ/year (0.46 Mtoe) from 2000 onwards. Another part of the plan allows individual towns and communities to adopt, if they wish, biomass technologies on their own initiative.

The *1991-1995 Centralised Biogas Plants Follow-Up Programme* shows that while public subsidy of initial investment costs has been steadily reduced from 30-40% in the 1980s to a maximum of 20% in the 1995, exemption of biomass from energy excise taxes remains a clear precondition to further development of this technology. Plans for development and demonstration of advanced biofuel technologies, in particular biomass gasification, are a major priority area within the biomass sector. Seven pilot plants are presently under construction or in operation. Two further inter-Ministry groups in the areas of solid biomass and liquid biofuels have also been set up.

Since 1992 there have been two *Solar Action Plans* (1992-94 and 1995-97), although capital subsidies for installation of hot water systems have been in place since 1979. Figures revealing the scale of possible penetration rates (though these are not targets) are given in the 1995-97 *Solar Plan,* which aimed to:

■ increase the installation rate of new solar collector systems from 1994 rates of around 2,500 plants per year to around 5,000 new systems per year by the end of 1997 (bringing total collector area of around 0.04-0.07 million m^2 at the end of 1997);

■ obtain a contribution to energy supply by active and passive solar of 0.2 Mtoe and solar cells 300 GWh by 2005 (around 1% of annual electricity consumption).

These initiatives were promoted via a direct subsidy, with the total budget of 30 M DKr annually. Subsides for active systems will be phased out in line with their deployment, with the balance of funding shifting in step to passive and solar cell systems. Recent collaboration between the government and the gas distribution companies has aided the installation of an increasing number of solar water heaters in combination with natural gas heating in dwellings. The demand from the gas companies of large quantities of solar water heaters has helped reduce the production costs per heater.

Trends in renewable energy R&D expenditure have shifted from wind (42% in 1990 down to 37% in 1996) and solar (from 18% in 1990 to 13% in 1996) towards biomass applications (rising from 28% in 1990 to 49% in 1996) producing electricity and heat. Overall support for renewable energy R&D projects has risen from 23.9 M DKr in 1990 to 36.5 M DKr in 1996.

STATUS OF RESOURCE EXPLOITATION

Renewable energy use more than tripled in the period 1970-1990, with developments accelerating recently, and likely to continue doing so as a result of the 2005 target. Renewable energy use has increased over the 1990s and contributed 6.9% (or 1.58 Mtoe) to Denmark's TPES in 1996, largely from biomass and wastes with the remainder mainly wind, solar and heatpumps. Electricity from biomass, wastes and wind generated 2393 GWh of electricity in 1996 (4.5% of 1996 total electricity generation). Denmark has a higher proportion of its electricity met from intermittent electricity sources (wind) than from any other country (although the proportion is significantly smaller if the perspective is that of the broader Nordic market in which Denmark participates). This proportion is expected to grow to approximately 15% of electricity by 2030. In 1996, wind power contributed 51% of the total electricity supplied from renewables and waste, biomass 17% and municipal waste 32%.

The use of heat derived from renewables provided 0.8 Mtoe in 1996: approximately 0.5 Mtoe in TFC (3.3% of TFC), with a similar amount generated and distributed as heat sold, mainly from municipal waste. Plans are for renewables to contribute around 10% to TPES by 2000.

Biomass

A number of utility-owned biomass plants (fired by wood, woodchips or municipal solid waste) have been in operation since 1989. ELSAM decided in mid-1994 to convert one of its 150 MW coal-fired units at Studstrupværket to co-fire straw, which will account for 10-20% of total input. However, conversion is costly, e.g. where additional gas cleaning is needed. Over the last few years, additional biomass units have been constructed, and (as outlined above) a certain amount of biomass use for electricity generation is mandated in the *Biomass Agreement*. In addition to these initiatives, a number of new demonstration biomass-fired plants have been planned or constructed, with over 200 MW 'biomass equivalent'[33] capacity expected to be commissioned by the end of the century.

Helped by these initiatives, biomass use has been growing steadily over the last few years in all applications (direct use, electricity generation and heat production). Biogas use has also increased sharply. In 1996, electricity production was 418 GWh (double the level of 1993), heat production was 10.9 PJ (up from 8.8 in 1992) and direct use was 509 ktoe – more than double its level in 1990. Continuing growth for biomass is planned and will be needed in order to meet the medium term target.

---

33 Calculated by evaluating a 100 MW unit firing 10% biomass as 10 MW biomass equivalent.

Waste

Over a third (38%) of renewable TPES is produced from municipal waste, and energy from waste combustion increased by almost 70% between 1990-1996. *Energy 21* aims for energy from waste to become increasingly important by 2005.

Wind

Wind power expanded rapidly in the late 1980s, growth slowed slightly in the early 1990s, and accelerated again from 1995 (50 MW of installed capacity in 1985 grew to 460 MW by 1992, 620 MW by 1995, 791 MW by 1996 and over 1GW by the end of 1997). This expansion was partially helped by the technological breakthrough brought about by variable speed turbines. Most of this expansion has been in the non-utility sector which accounts for 588 MW of the 791 MW 1996 installed capacity (of the total 172 MW additional capacity installed in 1996, 23 MW were installed by the utilities). Generation has continued to grow and reached 1217 GWh in 1996, 2.3% of total electricity generation. Installed turbines are now almost exclusively 600 kW or larger.

The 1500 MW target for installed onshore capacity by 2005 is around half the European Commission estimate of the technical potential. Since an agreement was reached between the Danish Ministry of Energy and Environment and the utilities to expand offshore wind capacity as well (750 MW by 2008), expansion of wind electricity will continue. The long-term target for total installed capacity (on-shore and off-shore) is 5.5 GW before the year 2030.

Solar

Only a small number of households (around 1-2%) have solar heating. The total use of solar energy in Denmark is very small (6 ktoe in 1996), but should grow following the recent installation of around 7000 combined gas/solar heaters. Generation from PV systems should also start soon, if the target for 2005 is to be met.

Hydro, Wave

Denmark has a very small (10 MW) hydro resource producing around 19 GWh in 1996. Potential markets for wave power are relatively more promising in the long term – two offshore 45 kW converters have undergone trials.

Geothermal

Plans to utilise Denmark's low enthalpy geothermal energy resources via a new demonstration plant have been put forward under *Energy 21*. One research plant is already in operation at Northern Jutland and produced 32 TJ of heat in 1996.

Heat Pumps

The net output of heat from heatpumps grew rapidly in the 1980s, and has been climbing slowly since 1992. The contribution to Denmark's TPES was estimated at 3141 TJ in 1996.

Table 2

Trends in renewable energy supply and use

	Unit	1990	1992	1995	1996	2000	1990-1996 (%)[1]	1996-2000 (%)[1]
Renewable TPES (excl. hydro)	**ktoe**	**1132**	**1309**	**1479**	**1579**	**2045**	**5.7%**	**6.7%**
Percentage of TPES	%	6.2	6.8	7.3	6.9	10.4		
Geothermal	ktoe	1	1	1	1	0	-6.5%	n.a.
Solar, Wind, Wave, Tide	ktoe	55	82	106	111	217	12.4%	18.3%
Biomass and Wastes[2]	ktoe	1076	1226	1371	1467	1828	5.3%	5.7%
- Biomass	ktoe	717	813	796	861	n.a.	3.1%	n.a.
- Wastes	ktoe	358	414	575	606	n.a.	9.2%	n.a.
Renewable electricity generation (excl. hydro)	**GWh**	**669**	**1133**	**2093**	**2393**	**6243**	**23.7%**	**27.1%**
Percentage of total generation	%	2.6	3.7	5.7	4.5	16.1		
Geothermal	GWh	0	0	0	0	0	n.a.	n.a.
Solar, Wind, Wave, Tide	GWh	610	915	1174	1217	2353	12.2%	17.9%
Biomass and Wastes[2]	GWh	59	218	919	1176	3890	64.7%	34.9%
- Biomass	GWh	n.a.	102	324	418	2670	n.a.	59.0%
- Wastes	GWh	n.a.	116	595	758	1220	n.a.	12.6%
Renewable TFC (excl. hydro)	**ktoe**	**203**	**502**	**511**	**538**	**507**	**17.6%**	**-1.5%**
Percentage of TFC	%	1.4	3.4	3.3	3.3	3.4		
Geothermal	ktoe	0	0	0	0	0	n.a.	n.a.
Solar, Wind, Wave, Tide	ktoe	2	3	5	6	11	17.0%	16.0%
Biomass and Wastes[2]	ktoe	201	499	506	532	496	17.6%	-1.7%
- Biomass	ktoe	201	499	490	509	496	16.8%	-0.7%
- Wastes	ktoe	0	0	15	22	0	n.a.	n.a.
Hydro TPES	ktoe	2	2	3	2	3	-5.7%	16.4%
Hydro electricity generation	GWh	27	28	30	19	30	-5.7%	12.1%
Percent of total generation	%	0.1	0.1	0.1	0.0	0.1		

Notes:

1. Annual Growth Rate
2. Including Animal Products and Gases from Biomass

FINLAND

OVERVIEW OF RENEWABLE ENERGY POLICY

Finland's lack of indigenous fossil fuels makes it a significant energy importer. Finland's goals of ensuring a secure, safe and environmentally acceptable energy supply have resulted in a commitment to develop diversified energy production capacity. Renewable energy promotion is therefore one of the main options to increase energy self-sufficiency open to Finland. The Government has recognised that promoting renewable energy also serves regional employment and social objectives. Finnish efforts to promote renewable energy focus on bioenergy and wind power. A Biomass Programme aims to increase use of bioenergy by 30% (1.5 Mtoe) from 1993 to 2005, and the government also aims to have 100 MW of wind capacity installed in the same year.

Economic and fiscal incentives and R&D are the two central policy approaches used to promote renewable energy in Finland. However, changes in Finland's Value Added Tax (VAT) system in order to harmonize it with the EU resulted in a higher VAT rate for biomass, which was previously exempt. Market reform in the domestic electricity sector, and increased integration into the Nordic electricity market is likely to have an effect on future renewable electricity developments: even the smallest consumers will be able to choose their electricity supplier, and this could help support Finland's wind and possibly bioenergy programmes (if consumers can choose "green power"). However, there is no guaranteed market for renewable electricity.

POLICIES

Finland imports a significant proportion of its TPES: there was a 46% ratio of indigenous production to TPES estimated in 1997. This is due to a lack of indigenous oil, natural gas and coal resources. Finland's domestic energy resources are limited to nuclear power, renewables and peat (which is not considered renewable by the IEA). The future development of Finland's indigenous biomass resources has been shaped significantly by commitments made in the government's May 1997 *Energy Strategy*. This strategy reinforces a long-standing Finnish commitment to increased use of bioenergy and other indigenous forms of energy.

In addition to the goals of energy security and environmental protection, regional employment and social benefits are clearly cited as key drivers for the overall renewable strategy. Increased use of renewable energy plays a prominent role in the country's climate change response plans, where the goal is towards the short-term stabilisation of greenhouse gas emissions. Supply-side climate change response options are limited to the development of renewables and replacing the use of coal with natural gas. Use of nuclear energy has been mentioned in the government's new strategy as an option for electricity production.

There are no targets for total renewable energy use. However, the Finnish government decided on 7 April 1994 upon an action programme for the promotion of bioenergy that aims to increase the use of bioenergy at least one quarter by year 2005. The programme aims to increase use of bioenergy by 30% in 2005 (compared with 1993). This is equivalent to an increase in bioenergy production of around 1.5 Mtoe per annum. The Government's approach to bioenergy subsidies, taxation and exemptions is explicitly linked to regional employment and social issues. The government also aims to have 100 MW of wind capacity installed by 2005.

In 1998, the Finnish share of the EU's -8% emission reduction commitment agreed under the Kyoto Protocol was confirmed under the EU's burden-sharing agreement at a stabilisation at 2010 compared to 1990 (as had been provisionally agreed pre-Kyoto). The Finnish climate change response programme[34] outlines the policies and measures it is undertaking to reach this goal, and also outlines domestic measures to enhance carbon reservoirs and sinks.

Energy policy is the responsibility of the Energy Department in the Ministry of Trade and Industry (MTI). The Finnish Technical Development Centre (TEKES), part of the MTI, aims to strengthen R&D activities and improve project management and cross-fertilisation with non-energy activities.

Fiscal measures and R&D are the two central policy approaches used to support market deployment and commercialisation of renewable energy in Finland. There are no regulatory measures or other related commitments. The *Electricity Market Act,* which entered into force in June 1995, allows all electricity producers to seek customers throughout the Nordic electricity market. This could allow interested parties to opt for "green power", if available.

Reduced fiscal incentives towards renewable energy, notably the removal of special VAT exemptions on bioenergy sources, came into effect from 1 January 1995 following requirements to harmonise with EU policy. This, and the 1997 energy tax system reform, have resulted in the favourable tax treatment of biomass inputs to electricity generation being removed: from 1997, tax has been levied on electricity output (irrespective of the fuel used for generation) rather than on the fuel inputs. However, a refund system was also introduced to compensate electricity generation for the reduced tax advantages: a tax refund equal to the electricity tax

---

34 *Finland's Second Report under the Framework Convention on Climate Change,* Ministry of Environment, Helsinki, 1997

is available for electricity from wind, wood, wood-based or small hydro. The need for further significant electricity supply capacity (estimated at 4800 MW above 1994 capacity by 2005) provides significant potential for the development of renewables, particularly biomass-based electricity generation. Impending decisions on how to expand base-load capacity will have an important impact on the prospects for renewable energy development.

However, the 1997 tax reform maintained the incentives for biomass-based heat generation, as taxes on heat are based on the net carbon emissions from input fuels, and are zero for renewable energy sources. Previously, the carbon/energy tax was based 60% on the carbon content of the fuel and 40% on the energy content. In addition, renewable energy sources are still subject to favourable tax treatment and direct bioenergy investment support to assist deployment and commercialisation.

For investments in new energy technologies, maximum subsidies are 30% of eligible cost, except for wind, where maximum subsidies are 40%. For energy technology studies, the maximum rate of subsidy is 50% of the amount of eligible cost. Investment incentives are available for commercial PV systems up to 30% of installation cost, although no incentives are available for small private systems, including households.

The development of bioenergy technologies is also given a high priority within energy R&D. Public expenditure on R&D is normally about 60 M FIM[35] per year. R&D finance is channelled mainly through the Research Programme for Bioenergy. PV and solar heating and cooling technologies are supported through construction and monitoring of demonstration plants. R&D expenditure for 1995-1998 on solar and wind is 120 M FIM. A set of eight national energy technology programmes was launched in 1993 by TEKES. Four programmes – Liekki 2, Bioenergia, NEMO 2 and SIHTI 2 – are directly related or have strong links with the promotion of renewable energy. Around one sixth or 2000 M FIM per year of total R&D expenditure is directed towards TEKES activities, of which 200 M FIM supports the energy sector. The funds for renewables is about 60 M FIM.

Long-term commitment to bioenergy has already produced some notable results. In Finland, wood accounts for a higher percentage of energy use than in any other industrialised country. CHP plants have been built in most cities and many industrial complexes where financially feasible. The biggest hindrance to such development is high costs relative to competing fuels. Wood fuel is used in the wood-processing industry and in the supply of energy to localities that have a wood-processing industry. With present technology and current public support, wood use can be economic in other sectors. The use of fuels made from field-grown biomass has also been investigated, but its use has not become more common.

The Advanced Energy Systems and Technologies research programme (NEMO2) is one of the energy research programmes of TEKES for the period of 1993-1998. The

---

35 In 1997, 1 US $ = 5.187 FIM

main emphasis is on solar energy and wind power. The total budget is about 120 M FIM.

Wind is also promoted directly by a Government programme. In its energy-policy report to Parliament in the spring of 1992, the government expressed its intention to draw up a long-term programme for expanding the wind power capacity in Finland. Finland's existing wind power capacity is around 11 MW. A MTI target for wind power in 2005 is 100 MW producing 0.25% of total estimated electricity requirements. Significant capacity growth in wind was not expected without such a systematic programme. The main objectives relating to support of wind energy projects are to reduce production costs, which at the moment are around double conventional generating costs, and to demonstrate the technology. It is recognised that specific targets in this area need to be set by government as part of a programme monitoring strategy.

A range of measures are in place to address non-financial barriers to the development of renewable energy technologies. Finland participates in the EU/ALTENER programme and MTI is developing a bioenergy information programme. Information and publicity initiatives are used to help overcome local planning problems. Representation of the energy utilities on the board of the renewable energy technology programmes is another supporting measure, injecting a utility view into renewables promotion.

A large demand for information about the promotion of use of renewables has led to a number of industry associations being established. These associations distribute information and collect statistics; the wind power, wood power and bioenergy associations are the largest. Efforts to promote further information and training on renewables are being intensified. The Ministry of Agriculture and Forestry and MTI are cooperating in their respective fields of competence, and particular attention will be paid to promoting increased use of wood fuel in the pulp industry.

There are few barriers to development of renewables, except for certain environmental constraints on wind power. Co-ordination between federal and local planning processes is not seen as a barrier. Finland has also supported two projects to promote penetration of renewable energy technologies overseas, including a study on the possibilities for the use of peat and wood fuels for electricity and heat supply in Kostamus (former Soviet Union).

STATUS OF RESOURCE EXPLOITATION

Finland's non-hydro renewable supply in 1996 was 5.26 Mtoe or 16.7% of TPES, compared to an IEA average of 3.9%. The vast majority (>95%) of non-hydro renewable energy was made up of biomass, but wastes and wind were also used to a small extent for energy purposes. In addition, hydropower generated 11.9 TWh in 1996, and accounted for 17% of total electricity generation in the same year. The largest user of biomass energy was industry, with residential and electricity/heat production accounting for the remainder. A small quantity of heat was generated from biomass and municipal waste sources.

Biomass

Biomass is an extremely important energy source in Finland, and accounted for over 16% of total energy supply in 1996: more than four times the IEA average. Black liquor and vegetal waste accounted for 2.6 Mtoe and 1.3 Mtoe respectively in 1996. The majority of biomass use is in industry: black liquor provided a third of the energy needs of the pulp and paper industry, which was the largest industrial user of this energy source. This industry obtains large quantities of wood-based fuels at a competitive price in conjunction with its procurement of raw materials. Significant quantities of black liquor are also used in wood production and in the chemical industry.

Vegetal wastes were mainly used for electricity and heat production (although some were also used in industry). Electricity production from biomass is often from multi-fuelled boilers, including in fluidised bed systems, which can combust moist biomass efficiently. Biomass is also used for space heating, and wood is estimated to provide 28% of space heating in the existing building stock. A 1996 survey of energy use of wood revealed 78% used by industry and 18% for space heating in the domestic sector.[36] The total technical potential of biofuels has been estimated at 10-12 Mtoe.[37]

Waste

Municipal and industrial waste together accounted for 171 ktoe of Finland's energy supply in 1996. Industrial waste is used for electricity and heat production as well as in industry. Municipal waste was used solely in CHP units.

Wind

The use of wind energy is currently very small in Finland. Wind electricity was first reported in 1993, at 4 GWh, and grew to 11 GWh by 1996. Growth will need to

---

36 Technical Research Centre of Finland research note 1559.

37 Finnish National Communication on Climate Change.

continue rapidly if the government is to meet its target of 100 MW by 2010: capacity in 1997 was 11 MW.

Hydro

In 1996, 2785 MW of installed hydro capacity generated 11.86 TWh of electricity (around 17% of total electricity generation). Total capacity, which is made up

Table 1
Trends in renewable energy supply and use

	Unit	1990	1992	1995	1996	2000	1990-1996 (%)[1]	1996-2000 (%)[1]
Renewable TPES (excl. hydro)	ktoe	4205	3840	4772	5258	5403	3.8%	0.7%
Percentage of TPES	%	14.6	13.9	16.5	16.7	16.6		
Geothermal	ktoe	0	0	0	0	0	n.a.	n.a.
Solar, Wind, Wave, Tide	ktoe	0	0	1	1	3	n.a.	33.4%
Biomass and Wastes[2]	ktoe	4205	3840	4771	5257	5400	3.8%	0.7%
– Biomass	ktoe	4176	3822	4759	5086	n.a.	3.3%	n.a.
– Wastes	ktoe	29	17	12	171	n.a.	34.3%	n.a.
Renewable electricity generation (excl. hydro)	GWh	n.a.	4950	6619	6154	6980	n.a.	3.2%
Percentage of total generation	%	n.a.	8.6	10.4	8.9	9.3		
Geothermal	GWh	n.a.	0	0	0	0	n.a.	n.a.
Solar, Wind, Wave, Tide	GWh	n.a.	0	11	11	35	n.a.	33.6%
Biomass and Wastes[2]	GWh	n.a.	4950	6608	6143	6945	n.a.	3.1%
– Biomass	GWh	n.a.	4950	6608	5889	n.a	n.a.	n.a.
– Wastes	GWh	n.a.	0	0	254	n.a	n.a.	n.a.
Renewable TFC (excl. hydro)	ktoe	3153	3014	3695	4022	4000	4.1%	-0.1%
Percentage of TFC	%	14	13.4	16.3	17.3	17.1		
Geothermal	ktoe	0	0	0	0	0	n.a.	n.a.
Solar, Wind, Wave, Tide	ktoe	0	0	0	0	0	n.a.	n.a.
Biomass and Wastes[2]	ktoe	3153	3014	3695	4022	4000	4.1%	-0.1%
– Biomass	ktoe	3153	3014	3695	3927	n.a	3.7%	n.a.
– Wastes	ktoe	0	0	0	95	n.a	n.a.	n.a.
Hydro TPES	ktoe	934	1299	1112	1020	1100	1.5%	1.9%
Hydro electricity generation	GWh	10859	15107	12925	11860	12793	1.5%	1.9%
Percent of total generation	%	20.0	26.2	20.2	17.1	17.0		

Notes:

1. Annual Growth Rate

2. Including Animal Products and Gases from Biomass

predominantly of large hydro plants, has grown only slowly during the 1990s, and is projected to expand 65 MW by 2000 from its level in 1996. Small hydro developments account for approximately 45 MW, but could be doubled in "easily exploitable" sites, e.g. at an abandoned station or dam. Total hydro generation in 1996 was approximately 10% lower than in 1995 due to lower precipitation.

Solar

Finland has a limited solar energy potential because of its climate. National estimates of PV systems installed by end of 1996 amounted to 1.5 MW. The majority of this is in solar home systems, and further use of PV systems to electrify remote holiday cabins is a potential market.

Geothermal; Ocean/Tidal/Wave; Heat pumps

None reported.

FRANCE

OVERVIEW OF RENEWABLE ENERGY POLICY

French policy is to promote renewable energies when they are competitive, or close to being competitive, in order to contribute to energy security, increased employment or for environmental reasons. Until recently, this led to emphasis on the direct use of biomass rather than on increased renewable electricity generation. The production of biofuels for transport is far from being competitive but has nevertheless been launched for long term research purposes as well as to provide some subsidies to the agricultural sector. In 1996 the government launched its "EOLE" programme, which aims to increase the supply of large-scale grid-connected wind electricity to at least 250 MW by 2005. Electricity generation from other renewables and wastes is supported by a purchase obligation on the state electricity company (EDF), although these renewable buy-back rates are relatively low.

France is one of the few IEA countries to have a significant population (mainly in its overseas departments and territories) that are not connected to the main electricity grid. The higher cost of electricity supply to these areas would in theory make renewable electricity supply an economically attractive option, particularly as these sites have significant solar and wind resources. However, EDF is legally obliged to supply low-voltage electricity at equal rates to consumers wherever they are located in metropolitan France or in overseas departments and whatever the cost to EDF. The resulting sale of some electricity at prices lower than its production cost effectively removes a niche market for (independent) renewable electricity production, and is therefore at odds with the aim to promote renewables where they are competitive. However, it also provides an incentive for EDF to promote renewables in remote locations where such use would either provide electricity, or reduce demand for electricity (and therefore financial losses from electricity production, e.g. via solar hot water heaters).

Reported use of renewable energy and wastes in France amounted to 4.2% of TPES in 1996 – higher than the IEA average, due mainly to the contribution of biomass, by far the largest non-hydro renewable source. Although France's electricity supply mix is expected to become slightly more dependent on fossil fuels in the medium term, the majority of electricity is currently generated by non-fossil sources: 78% of electricity was generated by nuclear power and 13% by hydropower in 1996 even though this was a dry year (normal hydro contribution is nearer 15%). From a greenhouse gas perspective, there are therefore few incentives to promote growth in new renewable electricity (other than wind, which is judged to be close to economic viability) and the contributions from these sources to electricity supply are therefore expected to remain insignificant in the short term. This is reflected in the small renewable R&D budget: France spends the lowest proportion of its energy R&D budget on renewables among IEA countries. However, other environmental considerations, notably regarding waste disposal, have increased interest in electricity generation or heat production from waste, while employment and agricultural considerations provide an impetus for the development of biomass.

POLICIES

Given the extremely limited national production of fossil fuels, maintaining security of supply remains a key objective of France's energy policy. Emphasis is also placed on low cost energy and environmental protection. Although long-term emphasis on nuclear power continues, a 1994 Government-initiated national debate and consultation exercise on future plans for energy policy emphasised among other things the need to stimulate EDF's actions to promote renewable electricity supply. Following the recommendations from this "Souviron Report", the government initiated EOLE, a plan to promote grid-connected wind electricity, in 1996. Promotion of wood energy for heating is also being strengthened via a *Wood Energy Plan*.

France's National Communication on Climate Change[38] outlines the future development of renewable energy, and quantifies emission reductions that are expected to result from the promotional policies put in place. The effect of all policies aiming to promote renewable energies and waste are estimated in the second National Communication as reducing 0.845 Mt carbon in 2000 and 1.145 Mt carbon in 2020. The majority of these reductions are estimated to come from decreased methane emissions from waste. (However, the potential impact of increased wind electricity is not estimated past 2005).

Renewable energy policy is formulated by the Ministry of Industry and implemented through the Agence de l'Environnement et de la Maîtrise de l'Energie (ADEME). Government supports renewable energy in several ways, including direct funding of local and regional projects, joint EDF/ADEME agreements, financial incentives (such as favourable tax treatment for renewable energy investments, reduced VAT on renewable energy equipment, and premium buy-back rates for successful projects under the EOLE programme) and information/education programmes.

ADEME is co-directed by the Ministries of Industry, Research and Technology, and Environment gaining financial support mainly from earmarked taxes and, to a lesser extent, directly from Government. ADEME supports several renewable R&D

---

38 The most recent is the *Second National Communication of France under the Climate Convention*, Mission Interministerielle de l'Effet de Serre [part of the French Government], November 1997, Paris

activities including: energy production from municipal waste; biomass (mainly the supply and distribution of wood); regional development plans; and PV standardisation. A tax on municipal waste was introduced in 1993 to encourage energy recuperation from waste. It stands at 30F/ton in 1996, and will increase to 35F/t in 1997 and 40F/t in 1998.

Increased use of biomass, especially for heating apartment blocks, is supported by the *Wood Energy Plan*, initiated by the Ministry of Industry. The total budget for the plan is 215 MF (of which 74.5 MF is from the national government via the Ministries of Agriculture and Environment, and the remainder from regional, local or EU funds) and the plan runs between 1995-1998. The plan aims to create 500 additional jobs by 2000 and should also result in fossil fuel savings of 60 ktoe. In support of this plan, France's 1997 budget lowered the VAT rate of 5.5% on wood used for home heating. State and/or local authorities also make available subsidies to cover all or part of the feasibility studies, equipment needs, incremental investment cost (compared to competing systems) and training necessary for wood energy heating systems. Subsidies are allocated on a case-by-case basis.

The EOLE programme is the largest programme available for promoting renewable electricity. It is directed solely towards wind-generated electricity, and aims for 250-500 MW capacity by 2005. It is run in a similar fashion to the UK's NFFO, whereby the government (in co-operation with EDF and ADEME) launches a competitive bid process for a certain amount of capacity. Successful bids are chosen on cost grounds: the average in 1997 was 0.337 FF/kWh, ranging between 0.3 and 0.5 FF/kWh. EOLE aims to drive costs down to (a competitive) 0.25 FF/kWh by 2005. Two bidding rounds had been held by the end of 1997, with projects for over 77 MW (more than initially anticipated) entering successful bids. Under the EOLE programme, projects have to be between 1.5-8 MW capacity: the legal limit for independent power producers. A market for power from the successful bids is guaranteed, and will be bought for the rate determined at the time of bidding for 15 years. A further round of bidding for 100 MW (of which 25 MW is to be in France's overseas territories and departments) was initiated in early 1998. A similar policy to encourage installation of 10 MW of biomass electricity capacity and 10 MW of biogas electricity plants was also announced in February 1998.

Financial support is not currently available for grid-connected PV systems. However, from 1993 until the introduction of the FACE scheme, such systems benefitted from a subsidy equivalent to 25% of the capital cost: 10% from ADEME and 15% by EDF. This subsidy was not high enough for many PV systems to be built. The Amortisation of Electrification Costs (FACE) fund is a source of finance for investments in renewables and demand-side management in rural areas. The annual budget for FACE is 100 MF. The majority of funds are spent on photovoltaic systems in rural areas, and aim to reduce either grid extensions or grid strengthening, via reducing peak demand or increasing stand-alone generation capacity.

There is also a programme that was launched by the government and EDF in early 1996 to install 20,000 solar water heaters by 2000 in French departments. Almost 4,000 heaters had been bought within the first year of this programme.

Biofuels benefit from excise tax exemption of up to 2.3 FF/l for RME and 3.3 FF/l for ethanol. The Government estimates that this subsidy will cost 1.5 billion FF per year in lost tax receipts, although it will only result in limited reductions in CO_2 emissions. This programme is followed for agricultural reasons, as it is not cost-effective in terms of CO_2 reduction alone. The ambitious, long-term goal for the biofuels programme is to use a methyl ester and ETBE biofuel split to substitute for 10% of transport fuel demand (twice the goal set by the EU).

Tax credits are available for investments in renewable energy technologies in overseas department for small hydro, wind, biomass photovoltaic and solar thermal power schemes, whereby renewable energy investments by a company can be deducted against taxable profits. France also uses information and education programmes to promote renewable energy via renewable energy advice centres as well as publications on renewable energy, such as *Systèmes solaires*.

National government expenditure on renewables accounted for 1% of total energy R&D budget in 1996 (US$5.8m). This was the lowest reported proportion of any OECD country's energy R&D budget that is spent on renewable energy. The majority is spent on biomass, photovoltaics and geothermal. A scientific agricultural group of chemistry and energy (AGRICE) has been set up to co-ordinate different research paths for increased use of energy crops, notably biofuels for transport and heating. Funding from the European Union THERMIE programme was used in the early 1990s to stimulate development of wind energy in France.

STATUS OF RESOURCE EXPLOITATION

Reported non-hydro renewable energy use in France stood at 10.7 Mtoe in 1996, the largest in any IEA country except the US. Non-hydro renewables accounted for 4.2% of total energy supply, compared to the IEA average of 3.9%. In addition, hydropower accounted for 12.8% of total electricity production, equivalent to a further 2.2% of total energy supply. The majority of non-hydro renewable energy use was solid biomass (particularly wood), almost all of which is used for residential heating. Municipal and industrial wastes are also being used to generate increasing quantities of electricity and heat. Small amounts of geothermal heat, solar energy and, increasingly, wind are also used.

Biomass

The use of wood for heating is widespread in France, with more than 3 million households using wood to fulfill their main heating requirements, and a further 4 million using wood heating occasionally. A further 1.5 million homes are estimated to occasionally use wood for heating purposes. Solid biomass (mainly wood) contributed an estimated 8.8 Mtoe to France's total energy supply in 1996. The majority of this, 7.1 Mtoe, was used for heating purposes in the residential sector. A small proportion of total biomass use was used to generate 714 GWh of electricity in 1996.

France is one of the few IEA countries to give a relatively high priority to the development of biofuels (largely for agricultural reasons, as outlined above), with 190 ktoe bioalcohol and esters produced in 1996 – approximately double the quantity two years earlier. Estimations for 1998 indicated a production of approximately 300 ktoe. Ethanol is transformed to ETBE whereas esters are used replace a small percentage of diesel.

Waste

Both municipal and industrial wastes are used, in approximately equal amounts, for electricity generation, estimated at 1.4 TWh in 1996. District heat production from waste incineration is increasing, and was reported as 47,500 TJ in 1996. (The exact estimations of outputs from district heating schemes are difficult because most use more than one energy source). Development of energy from wastes is set to increase as legislation prohibits landfilling of household wastes after 2002. Recuperation of landfill gas for greenhouse gas mitigation purposes should also help to increase energy production from landfill gas, which has not been developed to a great extent until now.

Wind

Wind capacity is small, but has been expanding over the last few years, and will continue to do so at a rapid rate due to the commissioning of the plants under the EOLE programme. Capacity was 900 kW in 1992, 3.4 MW in 1995 and around 10 MW at the end of 1997. Generation stood at 9 GWh in 1996. Projects totalling 77.5 MW of capacity had submitted successful bids by the end of 1997, and in theory these turbines should be installed by the turn of the century, as developers are given a three-year period in which to construct winning bids. A further round of bidding for 100 MW was launched in early 1998, and another series of bids will be held before 2005, which should result in an increase in short-term plans for new capacity.

Hydro

The capacity of hydropower plants has been stable at 20.5 GW since the early 1990s. However, weather variations lead to significant variations in generation from year to year, (77.3 TWh in 1994 and 65.2 TWh in 1996), and generation is expected to remain at around 67 TWh to the end of the century. Large hydro dominates both renewable electricity and total hydro output, with small hydro (<8 MW) contributing approximately 10% of hydro's total.[39] The majority of large hydro potential is already exploited, although an additional 4 TWh/y mostly from small hydro projects has been identified. Limits on the development of small hydro sites are generally due to flow requirements under water use regulations. Independent small hydro producers benefit from a purchase price guarantee for 15 years.

Solar

400 000 square metres of installed solar collectors provide 17 ktoe of heat (largely for hot water in residential buildings and for swimming pools), and 4000 solar water heaters were installed in 1996. National estimates for the capacity of installed PV systems were 2.5 MW in 1996, when generation was estimated at 2 GWh. However, without further incentives for solar electricity, it is unlikely to take off in the medium term, except in remote, rural, metropolitan districts and in overseas departments.

Geothermal

Geothermal energy is exploited via 41 low enthalpy geothermal heat plants around Paris and 15 in the Aquitaine region, estimated to supply around 121 ktoe heat in 1996. Electricity production from geothermal energy is being explored via a joint German/French/UK geothermal project on hot dry rocks underway at Soultz in Alsace. However, even a successful first phase of the project would not be expected to lead to a significant contribution to electricity supply for at least 10 years. A small geothermal station that operated between 1986 and 1992 in Guadeloupe, producing a maximum of 20 GWh per year, reopened at the end of 1997. There are no plans for significant short-term expansion of geothermal energy use.

Tidal

France is one of two IEA countries with installed tidal power. A large scale tidal installation at La Rance (240 MW) delivers approximately 550 GWh/y. No expansion of tidal power is planned.

---

39 Source: Ministry of Industry

Table 1
Trends in renewable energy supply and use

	Unit	1990	1992	1995	1996	2000	1990-1996 (%)[1]	1996-2000 (%)[1]
Renewable TPES (excl. hydro)	**ktoe**	**9986**	**11083**	**10767**	**10740**	**n.a.**	**1.2%**	**n.a.**
Percentage of TPES	%	4.4	4.7	4.5	4.2	n.a.		
Geothermal	ktoe	128	130	121	121	n.a.	-0.9%	n.a.
Solar, Wind, Wave, Tide	ktoe	61	63	64	62	n.a.	0.2%	n.a.
Biomass and Wastes[2]	ktoe	9797	10890	10582	10557	n.a.	1.3%	n.a.
- Biomass	ktoe	8627	9816	8867	8843	n.a.	0.4%	n.a.
- Wastes	ktoe	1170	1074	1715	1715	n.a.	6.6%	n.a.
Renewable electricity generation (excl. hydro)	**GWh**	**2234**	**2258**	**2672**	**2651**	**n.a.**	**2.9%**	**n.a.**
Percentage of total generation	%	0.5	0.5	0.5	0.5	n.a.		
Geothermal	GWh	20	6	0	0	n.a.	n.a.	n.a.
Solar, Wind, Wave, Tide	GWh	571	579	577	556	n.a.	-0.4%	n.a.
Biomass and Wastes[2]	GWh	1643	1673	2095	2095	n.a.	4.1%	n.a.
- Biomass	GWh	608	673	714	714	n.a.	2.7%	n.a.
- Wastes	GWh	1035	1000	1381	1381	n.a.	4.9%	n.a.
Renewable TFC (excl. hydro)	**ktoe**	**7584**	**8529**	**7520**	**7520**	**n.a.**	**-0.1%**	**n.a.**
Percentage of TFC	%	5.2	5.4	4.8	4.6	n.a.		
Geothermal	ktoe	0	0	0	0	n.a.	n.a.	n.a.
Solar, Wind, Wave, Tide	ktoe	12	13	14	14	n.a.	2.5%	n.a.
Biomass and Wastes[2]	ktoe	7572	8516	7506	7506	n.a.	-0.1%	n.a.
- Biomass	ktoe	7572	8516	7506	7506	n.a.	-0.1%	n.a.
- Wastes	ktoe	0	0	0	0	n.a.	n.a.	n.a.
Hydro TPES	ktoe	4588	5836	6136	5604	5800	3.4%	0.9%
Hydro electricity generation	GWh	53348	67859	71346	65160	66950	3.4%	0.7%
Percent of total generation	%	12.8	14.8	14.6	12.8	12.1		

Notes:

1. Annual Growth Rate

2. Including Animal Products and Gases from Biomass

GERMANY

OVERVIEW OF RENEWABLE ENERGY POLICY

German renewable energy policy aims to exploit more fully the long-term potential of many renewable energy sources[40]. The government has put in place a number of measures to promote renewable energy, notably the *Electricity Feed Law* (EFL). This requires utilities to buy renewable electricity generated from third parties at prices between 65-90% of consumer electricity prices. These high buy-back rates have made renewable electricity generation a commercially attractive option, and has led to a significant increase in renewable electricity generation and capacity since the EFL's introduction in 1990. For example, wind capacity shot up from 2 MW in 1990 to 2075 MW by end-1997 (the highest wind capacity of any IEA country).

Given the success of the EFL in increasing renewable electricity, renewable energy promotion is now shifting towards the promotion of the direct use of biomass and to heat production. Many policy types are used to this end, including economic incentives, improved information flows and R&D. There are no national targets for renewable energy, although some individual policies may have quantified goals. However, institutional barriers to renewable energy development are significant, and although an attempt has been made to standardise licensing of renewable energy technologies, full co-ordination of renewable development within and between Länder (Federal States) has yet to be achieved.

In 1996, 1.1% of Germany's TPES was met from non-hydro renewable energy sources, compared to an IEA average of 3.9%. However, this is expected to grow to almost 1.2% by 2000. One of the reasons behind Germany's promotion of renewable energy is its environmental commitments, particularly regarding CO_2. However, despite the increased use of renewable electricity, there remains a large potential for increased renewable energy use.

POLICIES

German energy policy continues to be characterised by efforts to reconcile political, economic and environmental objectives. The Chancellor's declaration in April 1995 to reduce CO_2 emissions by 25% by the year 2005 relative to 1990 has resulted in a range of policies and measures for renewable energy promotion being been put forward: the estimated effects of these measures are shown in Table 1.

---

40 German Energy Policy is described and analysed in detail in the IEA's *Energy Policies of Germany, 1998 Review*.

Perhaps the most effective renewable energy promotional policy to date has been the *Electricity Feed Law* (EFL), which is at the heart of renewables promotion and guarantees a market for renewable electricity at very favourable buy-back rates. Other government promotional policies include economic incentives for increased use of solar and biofuels, R&D programmes, favourable treatment of renewables in building codes and information and training. Green pricing (a private initiative) is available in some parts of Germany. Renewables are also promoted at the Länder level.

Responsibility for the promotion of renewable energies rests with a number of different groups at the national and regional level. The Federal Ministry of Economics drew up the EFL and also ran a four year 100 M DM[41] funding programme for new renewable energy sources. Demonstration of "environmentally sound" hydro projects in the new Länder is supported by a 10 M DM German Federal Foundation for the Environment programme. R&D on selected demonstration projects is supported by the Federal Ministry for Education, Science, Research and Technology (BMBF). Responsibility for biomass promotion rests with the Federal Ministry of Food, Agriculture and Forestry.

The drafting, financing and implementation of programmes to promote renewable energies in Germany are strongly influenced by the federal structure of the country. This can have the disadvantage that measures to promote renewable energies are not always co-ordinated in an optimal manner and that information that is of relevance for private interests is not always available to the extent desired. On the other hand, regional corporations have significant flexibility in promoting renewable energies. There are no plans to strenghthen co-ordination of renewable energy technologies on the Federal level or among the Länder.

Despite a number of attempts by some German utilities to block or overturn the EFL, it is still in place and creates a guaranteed market and favourable prices for renewable electricity generated by non-utilities. These prices are set at a percentage of final user prices, depending on the type of renewable: wind and solar electricity are paid 90% of the average consumer end price, small hydro facilities and biomass/waste are paid 80%, and large hydro schemes are paid 65%. For 1998, these prices are 16.79, 14.92 and 12.12 Pf/kWh respectively. These prices are approximately 2% lower than those paid in 1997 because electricity prices have been decreasing in Germany (due in part to the phasing out of the coal subsidy, the Kohlepfennig).

However, the EFL has been revised in order to distribute the cost of renewable electricity production, as it currently places a significant, although unequal, financial burden on some utilities (especially those near the coast where the majority of wind turbines are situated). The April 1998 *Energy Law* caps at 5% (of total electricity distributed) the amount of renewable electricity that has to be distributed by an individual (municipal) utility. Once this 5% threshhold is

41   In 1997, 1 $ US = 1.734 DM

Table 1
Main Measures to Promote Renewables and their effect on CO_2 Emissions

Measures	Expected CO_2 reductions in 2000/2005 (kt)	Comment
Federal Support to Renewables	143	This 100 M DM programme 1995-1998 provides subsidies for renewables
EFL	4 863/6 484	This has been in force since 1991 and has been the driving force behind the rapid expansion of windpower.
R&D on the use of renewables	23	Solar and biomass are the focus of long-term R&D actions.
Support for Testing Wind Power Systems	562	This programme provides investment or output subsidies for wind systems, and is estimated by the government to cost 400M DM by 2007.
Support for geothermal	35	Government funds approximately 4M DM R&D on geothermal per year.
50,000 Solar Roofs	30/80	This measure, financed by the Deutsche Auchgleichsbank, provides capital subsidises for households.
Promotion of Photovoltaic Energy	3	This "1000 roof" programme resulted in over 2000 PV installations.
Solarthermie 2000	1/2	This measure promotes solar water heating in public buildings.

Source: 2nd National Communication[42]

surpassed, the cost of supporting additional renewable generation is passed on to the higher level (regional) utility. If a regional utility also crosses the 5% threshhold, the obligation to pay for the renewable electricity passes to the supra-regional utility, and once this is surpassed the obligation to pay will end. The German government does not expect this to happen before 2000, and plans to revise obligations for the purchase of renewable electricity before then.

As the premium rates laid out in the EFL only apply to non-utilities, utility generation from new renewables has been extremely limited to date. However, the 5% limit set on renewable electricity distribution may, paradoxically, increase utility

---

42 *Climate Protection in Germany*, Second report of the Government of the Federal Republic of Germany Pursuant to the United Nations Framework Convention on Climate Change, Federal Environment Ministry, April 1997.

activities in this area: if utilities are already close to the 5% limit and can produce renewable electricity cheaper than the amount they would have to pay a third party for, it would be financially advantageous to generate it themselves. This is true even if utility-generated renewable electricity is more expensive than electricity from its other plants such as coal or gas.

The 100M DM programme run by the Federal Ministry of Economics encourages increased use of renewables via capital subsidies (up to a capped limit which varies by technology). Particular emphasis is given to solar collectors and heat pumps, small hydro power installations, large wind turbines (450 kW to 2 MW), PV installations greater than 1 kW and biomass installations. Some applications, such as solar water heaters for swimming pools, are excluded.

The Deutsche Ausgleichsbank (DtA), a public bank, grants low-interest loans to specified projects. About 4.18 billion DM of loans was granted between 1990 and December 1997, mostly for wind energy (3.48 billion DM) followed by hydropower (305M DM) and biomass (268M DM) in the framework of the ERP-Environment and Energy-Saving Programme and a specific environmental programme which includes the already mentioned "50,000 Solar Roofs" for households.

Technology-specific promotion programmes exist for PV, wind, and solar thermal systems. A large scale demonstration programme for photovoltaics: the "1000 roofs programme" was started in 1991, providing subsidies for production costs of 60% in the former GDR and 50% in western Germany. The programme has been completed and about 2100 units with total peak generation power of 5.3 MW are now installed. A five-year measurement programme is ongoing to monitor and evaluate this large pilot project. Data to be recorded include the energy generated and fed into the grid, and the energy purchased from the utility. One hundred systems chosen at random will be equipped with more detailed data acquisition systems to record time-dependent data, such as meteorological data. "Solarthermie 2000" was launched in 1993 by the Federal Ministry for Research and Technology (BMFT, now BMBF) as part of its large-scale demonstration programme for the development of low temperature heat from active systems, especially in the former GDR.

A variety of specialist municipal advice centres distributed around the country provide information resources on the use of renewable energies and the availability of incentives to promote renewable energy use. A brochure providing information and contact points on renewable energies was published in 1993 by BMBF. The Economics Ministry also periodically publishes a detailed source-book on different aspects of renewable energy technologies.

The German Government promotes use of liquid biofuels for use in motor vehicles through tax exemptions amounting to DM 1.5/litre for rapeseed methyl ester (RME) sold in around 250 filling stations around the country. Because RME is nearly commercial with this tax exemption, and there are no major technical problems for its use in cars, no further support programme is deemed necessary.

A range of other smaller programmes support renewable energy. From 1 January 1995, amendments to the Ordinance on the Fee Schedule for Architects and Engineers provide new incentives for architects and engineers to incorporate the use of renewable energies in dwellings. Land set-aside bonuses are available if biomass crops are grown on set-aside land. Subsidies for renewable energy also exist at the Länder level.

Customers of the largest supra-regional utility, RWE, can also opt to pay for "green power". Under this scheme, customers can choose to pay for certain quantities of power (minimum 100 kWh/year) at more than double its normal price. In 1997, almost 766 000 kWh were paid for at this rate. RWE undertakes to match the ex-VAT funds received and invest in wind, solar and small hydro plants.

Despite the range of promotional measures in place, the lengthy, inconsistent, and sometimes complex local approval procedures have proved a significant barrier to renewable energy development. However, the January 1997 change to the *Federal Construction Law* (Baugesetzbuch) expressly establishes preferences for wind and hydropower systems and should therefore help to overcome barriers relating to the siting of such systems.

Expenditure on renewable energy R&D is high and has increased significantly in recent years: 1996 expenditure was $103m (up $10m from the previous year), or almost 30% of the government's total energy R&D budget. The majority of funds are devoted to solar applications, particularly PV, although significant expenditure is also allocated to wind energy. Biomass and geothermal energy receive a much lower level of funding. The 1994 budget for the 'Agency for Renewable Raw Materials' for R&D concerning renewable raw materials was 56M DM.

STATUS OF RESOURCE EXPLOITATION

Renewable energy contributed 1.1% to Germany's TPES in 1996 – significantly below the IEA average of 3.9%. However, non-hydro renewables generated 1.8% of Germany's electricity in the same year. This pattern of renewable energy use in Germany is markedly different to many other IEA countries: in many countries, renewables' importance in TPES is higher than that in electricity generation, reflecting the importance of the direct use of biomass. In Germany, although biomass constitutes the majority of renewable energy used, it is mainly used to produce electricity and heat.

Electricity generated from renewables increased to an estimated 9.7 TWh in 1996 or 1.8% of total generation. The amount of renewable electricity produced has

115

almost doubled since 1990, due approximately equally to the takeoff of wind power and increased production from wastes and biomass. Hydropower accounted for a further 4.0% of German electricity production in 1996.

Biomass

The majority of biomass used in 1996 was the direct use of solid biomass, mainly wood but some vegetal waste and black liquor, in the residential sector (1.2 Mtoe). Although no supply forecast for biomass is made by the Government, experts estimate that biomass could contribute up to 2.1 Mtoe within the next 5-7 years (i.e. back up to its level of 1990) from its estimated technical potential of almost 29 Mtoe/year.

Waste

The utilisation of municipal and industrial waste for power generation has risen steadily, and generation has risen from 4.3 TWh in 1986 to 6.8 TWh in 1996. Growth has been fastest for the use of industrial waste: electricity generated from industrial waste rose from 2.4 TWh in 1990 to 4.1 TWh in 1996. Total capacity figures are not reported to the IEA but are estimated at over 700 MW.

Wind

To date, the growth in wind capacity and generation in Germany has been the success story of increased renewable energy use in IEA countries over this decade. The extremely generous incentives available for wind power, outlined above, resulted in reported installed wind capacity surging from 2 MW in 1990 to over 2 GW (including some turbines of 1 and 1.5 MW) by the end of 1997. Estimates for generation in 1997 were 3.3 TWh[43], up from 2.0 TWh in 1996 and 1.5 TWh in 1995.

Capacity and generation from wind turbines is, however, likely to slow over the next few years. Firstly, public acceptance is becoming more of a problem, especially in areas with a number of wind turbines. This may limit total electricity generation to between 5-9.6 TWh/year by 2000, compared to an estimated technical potential of 14-83 TWh. Secondly, and probably more importantly, the changes to EFL has raised concerns amongst financiers and entrepreneurs about the profitability of such systems, and this is likely to reduce the number of new systems built.

---

43 See: http://www.uni-muenster.de/Energie/wind/veroeff/hga9798.html (in German).

Solar

Installed photovoltaic capacity has risen to more than 34 MW by the end of 1996. The "1000 Roofs programme" had resulted in 2100 PV systems with a total capacity of 5.3 MW$_p$. It is estimated that in 1994 around 3.5 GWh of the 5 GWh PV production were generated by this scheme. The government supports the principle of installing PV installations on noise protection walls along motorways. Financial support is available from the EFL and the electricity is fed into the public grid. The post-2000 potential for PV utilisation has been estimated at 40-77 GWh.

The "Solarthermie 2000" programme runs from 1993-2002 and aims for 10,000 m² of installed collector surface, for the provision of hot water in small and large scale systems for a cost of between DM 0.2-0.3 per kWh$_{th}$. The technical potential of such systems has been estimated at 130 ktoe/year by 2000 and up to 50 Mtoe/year in the longer term.

Hydro

Hydro power accounted for the majority of total renewable electricity production, standing at 4.0% of total electricity production and 0.5% of TPES in 1996, generating 22.0 TWh. The potential for increased hydro production is limited, with an estimated maximum potential of 25 TWh at the beginning of the next century. Small hydropower (<5MW) is encouraged by the EFL, and micro-hydro stations (<500kW) benefit from the 100M DM renewable support programme of the Ministry of Economics. The amount and share of hydropower is expected to rise slightly to the end of the century.

Geothermal

Germany utilises its low enthalpy geothermal resource in around 20 locations to provide water and space heating. Use of geothermal energy is growing steadily. In 1993, national sources report that 34 MW provided 7 ktoe heat, while in 1994, 40 MW produced 8.5 ktoe (although none was reported to the IEA). The government's R&D programme aims towards potentials for 1995 and 2000 of 50 and 80 MW respectively.

Heat pumps

The potential for recovery of ambient heat (by heat pumps) is estimated to amount to between 0.11-0.17 Mtoe/year within 5-7 years and up to 9.5-11.9Mtoe/year in the longer term. This compares to current utilisation of approximately 86 ktoe/year.

Table 2
Trends in renewable energy supply and use

	Unit	1990	1992	1995	1996	2000	1990-1996 (%)[1]	1996-2000 (%)[1]
Renewable TPES (excl. hydro)	ktoe	**4115**	**3045**	**3644**	**3921**	**4170**	**-0.8%**	**1.6%**
Percentage of TPES	%	1.2	0.9	1.1	1.1	1.2		
Geothermal	ktoe	0	0	0	0	0	n.a.	n.a.
Solar, Wind, Wave, Tide	ktoe	4	24	127	176	340	89.6%	17.9%
Biomass and Wastes[2]	ktoe	4111	3021	3516	3745	3830	-1.5%	0.6%
- Biomass	ktoe	2332	1311	1492	1521	n.a.	-6.9%	n.a.
- Wastes	ktoe	1780	1710	2024	2224	n.a.	3.8%	n.a.
Renewable electricity generation (excl. hydro)	**GWh**	**5076**	**6004**	**8778**	**9666**	**13270**	**11.3%**	**8.2%**
Percentage of total generation	%	0.9	1.1	1.6	1.8	2.4		
Geothermal	GWh	0	0	0	0	0	n.a.	n.a.
Solar, Wind, Wave, Tide	GWh	44	278	1480	2043	4556	89.6%	22.2%
Biomass and Wastes[2]	GWh	5032	5726	7298	7623	8714	7.2%	3.4%
- Biomass	GWh	222	294	687	803	1333	23.9%	13.5%
- Wastes	GWh	4810	5432	6611	6820	7381	6.0%	2.0%
Renewable TFC (excl. hydro)	**ktoe**	**2277**	**1231**	**1313**	**1313**	**1180**	**-8.8%**	**-2.6%**
Percentage of TFC	%	0.9	0.5	0.5	0.5	0.5		
Geothermal	ktoe	0	0	0	0	0		n.a.
Solar, Wind, Wave, Tide	ktoe	0	0	0	0	0	n.a.	n.a.
Biomass and Wastes[2]	ktoe	2277	1231	1313	1313	1180	-8.8%	-2.6%
- Biomass	ktoe	2277	1231	1313	1313	n.a.	-8.8%	n.a.
- Wastes	ktoe	0	0	0	0	n.a.	n.a.	n.a.
Hydro TPES	ktoe	1499	1489	1873	1888	1600	3.9%	-4.1%
Hydro electricity generation	GWh	17426	17397	21780	21957	18668	3.9%	-4.0%
Percent of total generation	%	3.2	3.3	4.1	4.0	3.4		

Notes:

1. Annual Growth Rate

2. Including Animal Products and Gases from Biomass

GREECE

OVERVIEW OF RENEWABLE ENERGY POLICY

The Greek Government's new Law No 2244/1994 revised the code for electricity production from renewables and conventional fuels, has facilitated regulatory requirements and has increased the financial attractiveness of renewable electricity generation. The Ministerial Decision No. 8295/1995 that was issued to support the implementation of the above Law includes general technical and economic terms of the Purchase Agreements to be drawn up between electricity producers and the Public Power Corporation (PPC), the state electricity company. The effectiveness of these legislative measures is clearly indicated by a significant increase in the number of applications by investors in the energy sector[44] since the end of 1996.

Additional tools for renewable energy production include national and European programmes: the 5 year long Operational Programme for Energy/Subprogramme for renewable energy of the Ministry of Industry, Energy and Technology; the Regional Operational Programme; the programme for the Development of Industrial Research, the National Programme for Research and Technology, the programme RECITE, and EU programmes THERMIE, JOULE and ALTENER. The Greek Government have also set out targets for renewable energy use by the year 2000.

POLICIES

Security of supply and the development of diverse indigenous energy resources are the central pillars of Government's strategic approach to energy. Legislation drafted in 1996 examines the possibility of establishing a "Board for Energy Planning and Control" to co-ordinate planning and monitoring of national energy policy. The impetus behind renewable energy promotion, as well as having a significant potential in Greece, is the government's aim of supporting modest demand growth in electricity and the high priority placed on reducing urban air pollution. Space constraints are unlikely to limit the exploitation of renewable energy.

Promotion of renewable energy in Greece is based on four types of measure at the national level: economic incentives (in the form of both capital subsidies and favourable buy-back rates); targets for future renewable energy use; information

---

44 Greek energy policy is described and analysed in detail in the IEA's *Energy Policies of Greece, 1998 Review*.

dissemination, and R&D. In addition, EU programmes such as ALTENER, JOULE and THERMIE are employed to increase renewable energy use. Policies used to promote renewable energy development are described in the Greek National Communication[15], which estimates annual CO_2 savings of 3.1 Mt from increased use of renewables.

Specific measures promoting renewable energy sources (RES) are provided by recent legislation (Law 2244/1994, Ministerial Decision 8295/1995). The main provisions of these are to:

■ Remove restrictions and liberate regulations for electricity production from RES, with a maximum capacity of 50 MW for independent producers;

■ Remove restrictions for the exploitation of small water falls;

■ Allow auto-(self) producers the possibility of compensating on equal terms their own production of electricity from RES and their consumption (net metering);

■ Limit the amount of bureaucracy involved in the issuing of the licenses required;

■ Define all basic elements of the new improved pricing system.

National policy is formulated by co-operation between the Ministry's Directorates for Energy Policy, Renewable Energy Sources and Energy Conservation, the General Secretariat of Research and Technology, the PPC (the State power company) and the Center for Renewable Energy Sources (CRES). PPC have also established a company which will manage its renewable electricity production schemes. CRES is the National Center for the Promotion of Renewable Energy and the Rational Use of Energy (RUE) and it carries out R&D activities in the sector. CRES operates mainly under the auspices of the General Secretariat of Research and Technology, but also under the General Secretariat of the Ministry, responsible for the energy sector.

There are four forms of Government financial support available for renewables (with the emphasis on renewable electricity production):

■ Law 2244/1994 set up a new pricing policy between PPC and independent or auto (self) producers, whose installations are connected with either the PPC's isolated grids on islands or with the PPC's interconnected system in the mainland.

■ Law 1892/1990 provides economic incentives (a subsidy of up to 45 per cent reaching 55 per cent in certain cases, according to location) for the promotion of various investments including renewable energy production.

■ Law 2364/1995 allows for 75% of renewable appliances for households, such as solar water heating systems, to be deducted from a person's taxable income.

---

45 *Climate Change: The Greek Action Plan*, Ministry for Environment, Physical Planning and Public Works, February 1995, Athens.

■ The *Operational Energy Programme* of the Ministry of Industry, Energy and Technology, which runs to 1999, allocates 20 M ECU of public funds in addition to the 120 M ECU of EU and private funds for the development of renewable energy sources.

In addition, implementation of European Commission programmes such as THERMIE, JOULE and ALTENER, provides a series of important actions for reinforcing future exploitation of renewable energy technologies to 1999 by increasing R&D, dissemination and promotional activities.

Under Law 2244/1994, the PPC is required to purchase all electricity from private renewable electricity plants for a period of 10 years (after which an extension may be possible). For auto (self) producers selling electricity to isolated grids, all excess renewable electricity supplied to the PPC's network is purchased by PPC at 70 per cent of the low voltage tariff. When renewable electricity is fed into an interconnected system, the excess kWh produced is purchased at 70 per cent of the tariff corresponding to low, medium or high voltage. For independent producers, the PPC's purchase price for the excess kWh is set at 90 per cent of the low voltage and of the low, medium or high voltage tariffs, for the island's isolated grids and for the mainland connections, respectively.

In accordance with the relevant provisions of the Law 1892/1990 and 2244/1994, there is no difference between the subsidies for private investors for small hydro, solar, wind or biofuel electricity projects. However, as the uptake of the different energy sources will vary, so will the extent of public support. The *National Communication* estimates that 15 of the 85 billion Dr will be met by public funds, as will 10 of the 15 billion Dr for solar water heater promotion, 1 of 4 billion Dr for geothermal heat and 18 of 70 billion Dr for biomass promotion.

The first call for the Renewable Energies Subprogramme of the *Operational Programme for Energy* resulted in 26 projects totalling 94 MW receiving funding by September 1997. These include 23 MW of wind turbines, 7 MW small hydro, 5 MW PV and 59 MWth biomass. In addition, 4000 square metres of solar thermal applications have been funded under this programme to date.

Two studies have been carried out to estimate the potential future renewable energy market. The first was sponsored by the Greek Ministry for the Environment, Physical Planning and Public Works, and laid out targets for renewable energy in 2000 (outlined in the *National Communication*). The second preliminary study was carried out by CRES, March 1994, in the context of a national plan for the market deployment of RES by the year 2005. The Ministry of Industry, Energy and Technology is in the process of evaluating the results of these two studies. In addition, the PPC has targets for the development of renewable energy sources as part of its 10-year development plan.

Renewable R&D is carried out by CRES, institutions of higher education and other organisations such as the National Centre for Physics Research "Democritus", the

Table 1
Renewable Energy Supply in 1996 and Targets for 2000

Source	Target (Ministry of Environment)		Target (PPC only)	
	1996 (ktoe)	2000 target (ktoe)	1996 PPC Capacity (MW)	2003 target (MW)
Biomass	888	733	n.a.	n.a.
Wind	3.11	68	24.2	36.8
Small Hydro (<10 MW)	10.8	15	1.3 (systems < 1 MW)	17 ("small" hydro)
Photovoltaics*	0.01	} 156	0.17	1
Solar Heat	112		n.a.	n.a.
Geothermal Heat	2.7	20	37.6 (MWth)	200-300
TOTAL	1016	992	63.3	255-355

* Grid connected or non-connected islands and regions.

Source: *Renewable Energy Sources: Statistics for 1996*, Report for European Commission (Eurostat), March 1998

Institute for Geological and Mineral Exploration, PPC, etc. CRES is preparing a detailed national plan for the market deployment of renewables up to year 2005. Furthermore, the Government has earmarked 75 per cent of 100 million ECUs allocated for national R&D funding for renewables. R&D expenditure in 1996 was $3.4m, mainly spent on wind, biomass and geothermal energy.

Regional energy offices and/or centres are to be established under law 2244/1994. Their task is to promote new energy technologies and to prepare local plans for the exploitation of the existing RES and RUE potential. According to this law, CRES will assume a co-ordinating role for all such organisations. The remaining legislative and improvement of land planning and environmental provisions are now being undertaken by a strong co-operation between the Ministries of Environment, Physical Planning and Public Works and Industry, Energy and Technology. Another remaining barrier for the further promotion of RES is the lack of information amongst the general public on available technologies and on available financing opportunities. THERMIE and ALTENER are geared towards eliminating or reducing the effect of these barriers.

Obtaining appropriate land space for the installation of new renewable energy systems, especially wind parks on the islands, is a significant factor inhibiting their further development. The limited ability of the PPC's network to integrate these new installations into the existing system of energy distribution is also a constraint.

STATUS OF RESOURCE EXPLOITATION

Non-hydro renewable energy use in Greece was estimated at 1 Mtoe in 1996, or 4.1% of TPES. Hydropower contributed an additional 1.5% to TPES in the same year. Over two-thirds of non-hydro renewable energy use is made up of biomass used for direct heating in the residential sector. Biomass use in industry accounts for the majority of the remainder, although solar heat is becoming increasingly important, and small amounts of geothermal heat, wind and PV electricity are also used. Expansion of renewable energy use could lead to the proportion of renewable energy use growing substantially by 2010: the European Renewable Energy Study (TERES) estimate that renewables could contribute up to 12% of TPES.

Biomass

Use of wood for cooking and heating in households accounted for an estimated 0.7 Mtoe in 1996. Biomass use in this sector has been stable over the 1990s. Biomass use in industry (generally the wood industry using wood waste, but also cotton ginning residues, almond shells and straw) was estimated at 150 ktoe in 1996. TERES estimates the energy potential for forestry and agricultural wastes at 4.5 Mtoe. A CHP plant which uses cotton ginning residues, that has a capacity of 2.1 MWth and 0.5 MWe is currently in operation at Davlia. Some interest has been shown for the production of biofuels (bioethanol) to be used in the transport sector, and some R&D has been started in this area (although there is currently no Greek production of transport biofuels).

However, the use of biogas from landfills is planned: a 240 kW plant was constructed in 1996 in northern Greece (although had not yet started operating by March 1998), and a further biogas-recuperation installation is planned, and six additional permits for such systems had been submitted by early 1998. The food industry uses small quantities of biogas for space heating purposes.

Wastes

No use of municipal or industrial wastes (other than biomass-based wastes outlined above) reported.

Wind

A significant potential for wind energy exploitation exists mainly in the Aegean Islands region. Current use of wind energy is from 27 MW capacity generating 38 GWh. During the period 1990-96, 26 MW of wind energy capacity was installed, mainly in the Aegean Islands. By April 1998, six installation licences for

49.9 MW of wind capacity were granted for the island of Crete. PPC also plan to install two wind farms of 5 MW each on Crete by 2000, and PPC is also continuing its wind/diesel demonstration and R&D programme on Kythnos and Astypalea. PPC, via the possibilities provided by law 2244/1994, also intends to co-operate with private investors to undertake installation of new wind parks. Wind capacity is therefore likely to grow sharply in the next few years.

Hydro

Total hydro output was 0.374 Mtoe or 4348 GWh in 1996, of which the vast majority is made up of large hydro systems. Capacity in 1996 was 2.2 GW (the same as in 1995), representing approximately 25 per cent of the total economically viable hydro potential. A further 600 MW is under construction, including a 160 MW plant at Acheloos river, and should be in operation by the year 2000. Three small hydro plants, with a total capacity of 13.3 MW, are included in the first period of the PPC's Plan (up to 1998), and 68 further applications for small hydro plants with a combined capacity of 113 MW are at various stages in the planning process. Generation from small hydro plants has doubled since 1990, but expansion of generation from large hydro plants has been even more rapid.

Solar

Solar energy is used almost exclusively for water heating, although only a fraction of Greece's estimated market for solar thermal applications has already been exploited. Greece had an estimated 2.2 million m^2 of solar water heaters in place in 1996 and aims to increase this by a further 1.3 million m^2 by the year 2000.

PPC has already installed PVs of 233 kWp, mainly on small or isolated islands in the Aegean sea. PPC will also install a new PV plant of 60 kWp on Sifonos island by the end of 1999, and plans to install a further 900 kWp by the end of 2003.

Geothermal

There are good prospects for developing one third of the 450 MW economically exploitable low enthalpy potential by the year 2000. Two high enthalply geothermal fields of approximately 150 MWe exploitable capacity have been identified in Milos and Nisyros islands. A 2 MW high enthalpy geothermal unit is already installed in the island of Milos, and PPC has launched an information campaign to facilitate public acceptance for the further development of geothermal energy. PPC also plans to install a total 10 MW plant at a medium enthalpy field in Lesvos, combining electricity generation and sea water desalination. In addition, PPC is examining the possibility of exploiting geothermal energy as part of a joint venture, e.g. with local authorities. In 1996, geothermal output was 3.6 Mtoe.

Table 2
Trends in renewable energy supply and use[1]

	Unit	1990	1992	1995	1996	2000	1990-1996 (%)[2]	1996-2000 (%)[2]
Renewable TPES (excl. hydro)	**ktoe**	**951**	**969**	**991**	**1007**	**660**	**1.0%**	**n.a.**
Percentage of TPES	%	4.2	4.2	4.1	4.1	2.2		
Geothermal	ktoe	3	3	5	4	0	6.0%	n.a.
Solar, Wind, Wave, Tide	ktoe	75	88	109	115	110	7.4%	-1.2%
Biomass and Wastes[3]	ktoe	873	878	877	888	n.a.	0.3%	n.a.
- Biomass	ktoe	873	878	877	888	n.a.	0.3%	n.a.
- Wastes	ktoe	0	0	0	0	n.a.	n.a.	n.a.
Renewable electricity generation (excl. hydro)	**GWh**	**1**	**11**	**43**	**47**	**145**	**90%**	**32.5%**
Percentage of total generation	%	0.0	0.0	0.1	0.1	0.3		
Geothermal	GWh	0	0	0	0	0	n.a.	n.a.
Solar, Wind, Wave, Tide	GWh	0	8	34	38	145	n.a.	39.8%
Biomass and Wastes[3]	GWh	1	3	9	9	0	44.2%	n.a.
- Biomass	GWh	1	3	9	9	0	44.2%	n.a.
- Wastes	GWh	0	0	0	0	0	n.a.	n.a.
Renewable TFC (excl. hydro)	**ktoe**	**951**	**969**	**988**	**1004**	**650**	**0.9%**	**n.a.**
Percentage of TFC	%	6.1	6.1	6.0	5.7	3.2		
Geothermal	ktoe	3	3	5	4	0	6.0%	n.a.
Solar, Wind, Wave, Tide	ktoe	75	88	106	112	100	6.9%	-2.8%
Biomass and Wastes[3]	ktoe	873	878	877	888	n.a.	0.3%	n.a.
- Biomass	ktoe	873	878	877	888	n.a.	0.3%	n.a.
- Wastes	ktoe	0	0	0	0	n.a.	n.a.	n.a.
Hydro TPES	ktoe	152	189	303	374	400	16.2%	1.7%
Hydro electricity generation	GWh	1769	2203	3529	4348	4350	16.2%	0.0%
Percent of total generation	%	5.1	5.9	8.6	10.3	8.8		

Notes:

1. Renewables data formally submitted on an annual basis by country's administration to the IEA have been supplemented with more recent estimates.

2. Annual Growth Rate

3. Including Animal Products and Gases from Biomass

HUNGARY

OVERVIEW OF RENEWABLE ENERGY POLICY

Hungary is the only IEA country to have recently undergone the transition to a market economy. As such it is distinct from other IEA countries, having had a deep economic decline and a consequent reduction in energy use and CO_2 emissions. Moreover, despite recent price rises, the cost of energy is still not market-based to all consumer groups.

Hungary's most promising renewable energy source is combustible renewables and waste, as the majority of Hungary's land area is either under agricultural cultivation (almost 70%) or wooded (18%). However, the level of renewable energy use in Hungary is low, supplying 0.9% of TPES in 1996, almost all of which was combustible renewables and waste. This is due both to the lack of other large-scale renewable energy sources, and to their cost in comparison to other fuels.

Reduction of CO_2 emissions – one of the prime reasons for the promotion of renewables in other IEA countries – is less applicable in Hungary, as their current level of CO_2 emissions are lower than in 1990. However, the contribution that increased renewable energy use can make to improvements in local air quality is a factor in Hungary's objective of increasing its domestic use of renewable energy.

POLICIES

The Hungarian parliament adopted a new energy policy in 1993 (Resolution 21/1993 (IV.9.)). The primary strategic objective is to maintain energy security and to improve energy efficiency, while limiting environmental impacts. This last objective includes a specific aim to increase the use of renewables to 5% of total energy supply by 2000. Renewable energy promotion is also included in the government's *Action Programme for Energy Conservation.*

Hungary's transition to a market economy, which has been taking place since the late 1980s, has resulted in significant structural changes. GDP, industrial production and energy use declined sharply at the beginning of the 1990s. Despite the slight turnaround since 1992/3, energy supply in 1995 was almost 20% below its 1987 high. Limiting increases in CO_2 emissions, which is one of the reasons behind many other countries' promotion of renewable energy, is not so important to Hungary as 1995 emissions were significantly below their 1990 level.

Three bodies have responsibility for developing renewable energy policy in Hungary: the Department of Energy (in the Ministry of Industry, Trade and Tourism), the Ministry of the Environment and Regional Policy, and the Commision on Sustainable Development. The Commission is led by the Ministry of the Environment and Regional Policy, and is responsible for the development of policies and measures related to sustainable development, including climate change. This Ministry is also responsible for the development of pollution-limiting legislation. The Ministry of Industry, Trade and Tourism is responsible for developing general energy policy and waste management targets. Energy prices are set by another body: the Hungarian Energy Office, which was set up in 1995.

Renewable energy in Hungary is promoted via economic incentives: indirect subsidies and interest-free loans for research and development activities. The primary reason for promoting increased use of renewables is to mitigate local pollution: over 13% of Hungary's land is at least "moderately" polluted. Much of this pollution is due to the use of indigenous low-quality, high-ash and high-sulphur coal in power generation.

Subsidies are available from certain funds, e.g. the central environment protection fund and the regional development fund that operate under the *Action Programme for Energy Conservation*. While these funds do not target only renewable energy systems, renewable energy applications are preferred.

Recent changes in the energy sector have included the privatisation of the electricity industry: the generation monopoly of the power utility MVM was ended in 1994. However, utilities still account for almost all electricity generating capacity. This lack of non-utility generating capacity is due to the demand for electricity contracting sharply since 1989 and also because electricity prices do not yet fully reflect market values, despite a series of price hikes in 1996 and early 1997.

Lack of market pricing in the electricity sector affects the potential for renewable electricity development: policies that other countries use to promote renewable energy, such as offering premium prices for renewable electricity via a levy on consumer bills, are unlikely to be politically acceptable. The relatively low level of electricity prices is also likely to hinder investment in renewable electricity technologies.

Unlike most other IEA countries, there are no basic R&D activities on renewable energy in Hungary. However, the incentives available for renewable energy can be applied to R&D activities, but any interest-free loans received would have to be paid back in case of project failure.

STATUS OF RESOURCE EXPLOITATION

Non-hydro renewable energy accounted for 0.9% of Hungary's TPES in 1996, and has been at this level since 1993. The majority of renewable energy use is the direct use of combustible renewables and waste (by-products from agricultural and forestry activity). There is significant potential for increased use of biomass and wastes, given the importance of the Hungarian agricultural sector. A limited amount of geothermal energy (hot water for heating) is used in Hungary, and there is little room for further expansion of this resource. The current use of hydro is negligible, and although there is potential for increased use of mini-hydro systems, Hungary possesses no large hydropower resources. Similarly, solar and wind resources are limited.

Biomass

Reported biomass use was 218 ktoe in 1996, a similar level to 1995 but lower than biomass use in the early part of the decade. The majority of this was used directly with a small amount also being used for heat generation. Biomass use could be expanded greatly if some of the currently uncultivated land was used to cultivate species that could be used for energy purposes (e.g. ligneous plants or crops from which transport biofuels could be derived).

Waste

No use of wastes for energy purposes is reported to the IEA.

Wind

The wind resource of Hungary is well documented, and iso-lines mapped. However, Hungary is an inland (and flattish) nation, and wind speeds are low: averaging 2.6m/s at ground level. While wind-speeds at 30-50m (turbine height) should be higher, these have not been mapped. There is currently no experience with wind-generated electricity in Hungary, nor are there plans to install any wind turbines.

Hydro

Some micro-hydro systems are in place in the northern and eastern parts of the country, although the power produced from these river systems may be seasonal due to variations in flow rates. Total installed capacity from micro-hydro is approximately 400 kW – this has remained stable for the past few years. Hydro power generation was 207 GWh in 1996 (0.6% of total generation from 48 MW capacity).

Table 1
Trends in renewable energy supply and use

	Unit	1990	1992	1995	1996	2000	1990-1996 (%)[1]	1996-2000 (%)[1]
Renewable TPES (excl. hydro)	**ktoe**	**365**	**320**	**221**	**218**	**990**	**-8.2%**	**46%**
Percentage of TPES	%	1.3	1.3	0.9	0.9	3.4		
Geothermal	ktoe	0	0	0	0	0	n.a.	n.a.
Solar, Wind, Wave, Tide	ktoe	0	0	0	0	0	n.a.	n.a.
Biomass and Wastes[2]	ktoe	365	320	221	218	990	-8.2%	46%
- Biomass	ktoe	365	320	221	218	n.a.	-8.2%	n.a.
- Wastes	ktoe	0	0	0	0	n.a.	n.a.	n.a.
Renewable electricity generation (excl. hydro)	**GWh**	**0**	**0**	**0**	**0**	**130**	**n.a.**	**n.a.**
Percentage of total generation	%	0.0	0.0	0.0	0.0	0.3		
Geothermal	GWh	0	0	0	0	0	n.a.	n.a.
Solar, Wind, Wave, Tide	GWh	0	0	0	0	0	n.a.	n.a.
Biomass and Wastes[2]	GWh	0	0	0	0	130	n.a.	n.a.
- Biomass	GWh	0	0	0	0	n.a.	n.a.	n.a.
- Wastes	GWh	0	0	0	0	n.a.	n.a.	n.a.
Renewable TFC (excl. hydro)	**ktoe**	**363**	**319**	**202**	**217**	**871**	**-8.2%**	**41.5%**
Percentage of TFC	%	1.7	1.8	1.2	1.2	4.1		
Geothermal	ktoe	0	0	0	0	0	n.a.	n.a.
Solar, Wind, Wave, Tide	ktoe	0	0	0	0	0	n.a.	n.a.
Biomass and Wastes[2]	ktoe	363	319	220	217	871	-8.2%	41.5%
- Biomass	ktoe	363	319	220	217	n.a.	-8.2%	n.a.
- Wastes	ktoe	0	0	0	0	n.a.	n.a.	n.a.
Hydro TPES	ktoe	15	14	14	18	20	2.5%	3.0%
Hydro electricity generation	GWh	178	158	163	207	200	2.5%	-0.9%
Percent of total generation	%	0.6	0.5	0.5	0.6	0.5		

Notes:

1. Annual Growth Rate

2. Including Animal Products and Gases from Biomass

Solar

Although no solar energy use is reported to the IEA, an estimated 20-25000m^2 solar hot water heaters are in place in Hungary.

Geothermal

No use of geothermal energy is reported to the IEA. However, Hungarian sources[46] estimate that approximately 80-90 ktoe of hot water is used within the agricultural sector and in nine town's district heating.

Heat Pumps

No use of heat pumps is reported to the IEA.

---

46 Ferenc Bohoczky, Department of Energy, *Renewable Energy Resources and their Utilisation in Hungary*.

IRELAND

OVERVIEW OF RENEWABLE ENERGY POLICY

One of Ireland's energy objectives is to produce as much of national energy requirements from indigenous sources as is economically possible. The government see development of renewable sources of energy as an appropriate strategy to meet this objective, as well as providing opportunities for rural development. Environmental considerations, specifically CO_2-free electricity production, are also important.

The contribution of non-CO_2 energy sources to the country's total primary energy requirements is low (partly due to the lack of large hydro resources). The main renewables sources used to date have been hydro and woody biomass, and to a lesser extent, wind power, solar, and geothermal. Although the current contribution from these sources is small, their energy use has been expanding rapidly, and policies currently in place should further increase the amount of energy supplied from renewable energy sources. The government has an ambitious target that 10% of electricity generating capacity should be from renewable energy sources by the end of 1999.

Government support has concentrated on renewable electricity generation, via the *Alternative Energy Requirement* (AER): a competitive tendering process whereby electricity from selected renewable energy projects is guaranteed a market, and Ireland's main programme for the promotion of renewable energy. There have been four AERs since the launch of the programme in 1994, with an aggregate target of 230 MW of renewables and CHP. Approximately 300 MW of renewable electricity-generating capacity have bid successfully for projects under AERs I , II and III. AER I guaranteed a "strike" price, and subsequent AERs guarantee the project bid price for 15 years. Wind, hydro, waste-to-energy and landfill projects have all provided successful bids (no solar project has been supported under this programme). Although interest in the programme has resulted in a greater-than-expected number of successful project bids, holdups in obtaining planning permission for renewable energy plants has slowed the development of installed renewable energy capacity. A 1996 review of renewable energy policy resulted in the government setting post-2000 targets for the development of wind, hydro and biomass generating capacity.

POLICIES

Promotion of renewable energy receives a high profile in Irish energy policy. Renewable energy is promoted in Ireland because they are indigenous energy sources, help energy diversification, reduce environmental damage of energy production and because they contribute to limiting the increase in CO_2 emissions. Ireland has agreed under the FCCC to minimise increases in CO_2 emissions in 2000 to no greater than 20% above 1990 levels. It's target under the EU burdensharing agreement, negotiated post-Kyoto, is +13% to 2008-2012. The 1993 *Irish National CO_2 Abatement Strategy* and the Irish National Communication refer to the role that wind, solar energies and energy crops could play in mitigating CO_2 emissions, and the proposals contained in the National Communication for a renewables tranche of electricity capacity have been carried out through the *Alternative Energy Requirement* programme.

The Minister for Public Enterprise has sole responsibility for renewable energy policy. An Energy Advisory Board appointed by the Minister for Energy provides advice on policy and programmes in the field of renewable energy resources, while responsibility for co-ordination and implementation of energy conservation lies with the Irish Energy Centre.

The AER[47] bids have encouraged the development of renewables: less than 8 MW of renewable energy generating capacity was in place prior to the start of the AER, and bids for more than 140 MW of renewable electricity plant were successful in AERs I and II (although not all of this capacity is likely to be built mainly due to delays in or refusal of planning permission). AER I was supported by an 15[48] M IR£ fund for grant assistance for project capital costs from the European Regional Development Fund and 70 M IR£ for long-term price support for successful projects. This latter sum will be passed on to the consumers by the Electricity Supply Board (ESB). The successful renewable electricity projects were chosen via a competitive bidding procedure, where like technologies compete with like (as in the UK's NFFO). There have been three AERs to date, the result of which are outlined below.

In AER I, prospective generators were invited to bid to sell electricity to the ESB under purchase power agreements (PPAs) for up to 15 years. All renewable energy projects under AER I were paid a "standard" price (6.1-6.4p/kWh during weekdays, and 2.4-2.5p/kWh at nights and weekends) that is indexed annually based on the Consumer Price Index. Prospective producers could also bid for a capital grant to ensure the recovery of project costs and a return on investment where the tariff revenue alone did not ensure this. However, not all capacity bids submitted a grant request, as the developers estimated that they could recover their costs through the purchase power agreements and the target renewable capacity was in fact exceeded without needing to disburse any capital subsidies.

In AERs II, III and IV[49], PPAs are also available for up to 15 years, but the prices paid will reflect the bid price of successful projects. A cap was set of 3.6p/kWh for AER

---

47 Although the government's definition of alternative energy includes CHP, discussion in this chapter will
    focus where possible on the renewables component only of the AER.

48 On average in 1997, 0.660 IR£ = 1 US $

49 AER IV is exclusively for CHP projects.

II (biomass/wastes) and 3.9p/kWh for AER III (5p/kWh for potential wave-based electricity projects). The actual price paid for electricity will be adjusted by season and time of day so that electricity produced at peak time will be paid more than electricity produced during off-peak hours. Successful bids from AERs II and III are expected to be operational by the end of 1999.

AER I resulted in successful bids for 73 MW wind, 12 MW landfill gas and 4 MW small hydro. AER II resulted in one successful bid of a 30 MW waste-to-energy plant. The target capacities for AER III are 90 MW wind (made up of 25 MW for wind farms under 5 MW and the remainder for wind farms of 5-15 MW), 7 MW biomass/waste and 3 MW small hydro. These were all surpassed, with successful bids of 137 MW wind, 17 MW biomass/waste (mainly landfill) and 4.4 MW small hydro.

Following the successful completion of the bidding and allocation procedure under AER I, a review of alternative energy strategy was initiated. This process included consultation with all interested parties, and a number of studies for the review were commissioned by the Department. The Government concluded that the AER was a significant step towards using Ireland's renewable energy resource and that the bids received for AER I indicated a willingness to invest in alternative energies. However, the government acknowledge that several players are involved in the development of renewable energy in Ireland, and that the concerns of each of these players will need to be addressed in the development of a successful renewable energy industry.

In particular, the lengthy planning procedures required and grid access for renewable-based electricity were listed as the most significant barriers to increased penetration of renewable energy. Both of these were addressed in the first AER following the review (i.e. AER III): one of the technical requirements for systems under AER III was for proposed projects to have secured or to "have good prospects for" planning permission. Third party access to the electricity grid is now allowed for renewable energy generators who wish to sell their green electricity direct to the public. However, the cost of access needs to be determined in the context of the transportation of the EU Directive.

Following this review of renewable energy policy, government targets for future renewable energy development have been announced (Table 1). These will be supported through the funds remaining from AER-1, and all projects will receive a grant of 65,000 IR£/MW installed.

The Government's general policy document "An action Programme for the Millennium" commits the government to promote pilot alternative energy projects and to publishing a Green Paper on environmentally sustainable energy sources. The AER programme will continue after 2000, with annual competitions expected.

The restructuring of the electricity utility ESB to create a competitive electricity market could impact the future viability of renewable projects not covered under

Table 1
Targets for renewable electricity generating capacity

	Issued/ revised	Target	Comment
"Alternative" electricity capacity	1994	75 MW by 1997	68.3 MW installed, February 1998.
Renewable electricity capacity (excl. CHP)	1996	10% of electricity capacity by 1999	Accelerated implementation/ commissioning of successful projects is needed if this target is to be met on time.
Wind capacity	1996	300 MW installed 2000-2010	Success dependent on future AER announcements.
Hydro capacity	1996	10 MW installed 2000-2010	Success dependent on future AER announcements.

the AER. Other regulatory measures that could influence renewables include the production of guidelines (from the Department of the Environment) for planning authorities on the development of wind farms. In December 1997, the government indicated that tax relief will be introduced for private investment in approved wind energy and biomass projects: this relief will apply to up to 50% of a project's cost (capped at 7.5 M IR£ per project and 10 M IR£ p.a. per company).

Renewables formed a significant percentage of government-funded energy R&D in 1992. National sources indicate that renewable energy projects receive up to one-third of current energy national R&D expenditure (although details are not reported to the IEA) and have also received significant assistance from the European Union programmes. While the predominance of funding for renewable R&D is European (65%), the state provides indirect support through universities and other institutions. In 1992, the majority of the total renewables R&D funds of 87,000 IR£ were spent on hydropower, with some spent on biomass. Only 3,000 IR£ was spent on solar energy and none on wind.

The Irish Energy Centre established a regional Renewable Energy Information Office in Bandon, Co. Cork in September 1995. This office will provide advice and information nationally on renewable energy as well as supporting the Irish Energy Centre's initiatives at a regional level.

STATUS OF RESOURCE EXPLOITATION

Non-hydro renewable energy provided 119 ktoe (1%) of Ireland's total energy supply and 0.3% of total electricity supply in 1996. Both are significantly below the IEA average. In addition, hydropower contributed a further 722 GWh (3.8% of total electricity supply) in the same year. Non-hydro renewables' contribution is made up almost exclusively by the direct use of woody biomass, landfill gas and a small quantity of wind-generated electricity. As plants from AER I, II and III come on stream, however, the proportion of renewable-generated electricity – especially from wind – should increase substantially.

Biomass

Wood accounted for all of Ireland's biomass use (109 ktoe) in 1996. All biomass is reported as being used directly, mainly in the industrial or residential sector. No biomass-generated electricity was produced in 1996, but 11.8 MW of the 12 MW biomass/waste projects successful under AER I were operational by mid-1997 (fuelled by landfill gas). The results of AER III and the biomass target outlined in the previous section should result in further capacity (probably landfill and/or other biogas) being on-line by 2000.

Waste

The successful AER II project was a 30 MW waste-to-energy plant that will use about 50% of Dublin's non-hazardous waste to produce electricity. However, this is unlikely to be in operation before 1999, and current use of waste for energy purposes is negligible.

Wind

Ireland's first demonstration wind farm in Bellacorrick was completed in 1992, with an installed capacity of 6.45 MW which generated 18 GWh in 1994. Total wind generation in 1996 was 1.2 ktoe. A further 73 MW bid successfully under AER I, although problems in the planning process means that not all of this will be commissioned. A further 137 MW submitted successful bids under AER III. By February 1998, Ireland's wind capacity stood at 36.15 MW, with approximately another 10 MW from AER I still to be completed. However, the interest in wind projects has exceeded government expectations, and the majority of future renewable energy development in Ireland is expected to come from wind farms – this is reflected in the renewable energy targets outlined in the previous section.

Hydro

Hydropower generated 920 GWh of electricity in 1994 or 5.5% of electricity production. This dropped to 722 GWh and 3.8% of total electricity in 1996, due to lower precipitation. The EC has estimated that up to 75% of viable hydro resources has been already developed; the remaining 25% consisting of smaller installations in more difficult sites, although there is scope to use existing capacity more effectively. Capacity is therefore expected to grow only marginally – 9 MW – between 1994 and 2000. Only 67 MW of the current 232 MW capacity is small hydro (< 10 MW). Exploitation of small hydro sites is encouraged by the AER programme, with 3 MW expected to be commissioned from AER III, and a further 10 MW in the first decade of the next century.

Solar

There is only one demonstration PV plant in Ireland. It has a peak power of 55.5 kW and produces about 20,000 kWh per year. The technology has never proved of commercial interest apart from minor uses in remote applications.

Geothermal

Some geothermal R&D projects were started in 1990 to utilise resources in Mallow, Co. Cork and evaluate the potential of deep boreholes in Munster and Leinster. Naturally heated ground water occurring in Dublin City Centre is used in several medium-sized buildings, and estimated to produce around 10 ktoe of useable heat. Geothermal energy use is not expected to grow by 2000.

Heat Pumps/Wave/Other

Ireland reports that heat extraction from heat pumps has risen slightly from 2 ktoe in 1990 to 3.5 ktoe in 1993. No forecasts for the future use of heat pumps is available.

The government wish to explore the possibility of wave-generated electricity, and indicated their willingness to support a pilot wave demonstration plant under AER III (subject to the bid price being below 5p/kWh) which will be completed by 1999. However, no bids met this criteria for AER III.

Table 2
Trends in renewable energy supply and use

	Unit	1990	1992	1995	1996	2000	1990-1996 (%)[1]	1996-2000 (%)[1]
Renewable TPES (excl. hydro)	**ktoe**	**2**	**91**	**166**	**119**	**203**	**102%**	**14.4%**
Percentage of TPES	%	0.0	0.9	1.5	1.0	1.6		
Geothermal	ktoe	0	0	0	0	10	n.a.	n.a.
Solar, Wind, Wave, Tide	ktoe	2	2	3	3	36	9.8%	85.8%
Biomass and Wastes[2]	ktoe	0	89	162	116	157	n.a.	7.9%
- Biomass	ktoe	0	89	162	116	n.a.	n.a.	n.a.
- Wastes	ktoe	0	0	0	0	n.a.	n.a.	n.a.
Renewable electricity generation (excl. hydro)	**GWh**	**20**	**25**	**52**	**62**	**640**	**20.8%**	**79.2%**
Percentage of total generation	%	0.1	0.2	0.3	0.3	3.0		
Geothermal	GWh	0	0	0	0	0	n.a.	n.a.
Solar, Wind, Wave, Tide	GWh	20	25	36	34	422	9.2%	87.7%
Biomass and Wastes[2]	GWh	0	0	16	28	218	n.a.	67.0%
- Biomass	GWh	0	0	16	28	n.a.	n.a.	n.a.
- Wastes	GWh	0	0	0	0	n.a.	n.a.	n.a.
Renewable TFC (excl. hydro)	**ktoe**	**0**	**89**	**158**	**109**	**105**	**n.a.**	**-1.0%**
Percentage of TFC	%	0.0	1.1	1.9	1.3	1.0		
Geothermal	ktoe	0	0	0	0	10	n.a.	n.a.
Solar, Wind, Wave, Tide	ktoe	0	0	0	0	0	n.a.	n.a.
Biomass and Wastes[2]	ktoe	0	89	158	109	95	n.a.	-3.4%
- Biomass	ktoe	0	89	158	109	n.a.	n.a.	n.a.
- Wastes	ktoe	0	0	0	109	n.a.	n.a.	n.a.
Hydro TPES	ktoe	60	70	61	62	69	0.6%	2.7%
Hydro electricity generation	GWh	697	817	713	722	803	0.6%	2.7%
Percent of total generation	%	4.9	5.2	4.0	3.8	3.7		

Notes:

1. Annual Growth Rate

2. Including Animal Products and Gases from Biomass

ITALY

OVERVIEW OF RENEWABLE ENERGY POLICY

Decreased reliance on imported fuels, which account for approximately 80% of Italy's energy requirements, and commitments under the FCCC are the main reasons behind Italy's promotion of renewable energy. The existence of several isolated dwellings for which grid-connected electricity is an expensive option, and favourable climatic conditions for PV and wind also provide opportunities to increase the takeup of renewable energy technologies in niche markets. Although the 1988 *National Energy Plan* and its quantitative energy policy objectives regarding renewable energy are no longer being followed, the general thrust to increase the penetration of renewable energy remains (specifically renewable electricity, although to a lesser extent solar heat and direct biomass use).

Encouragement of renewable energy started in earnest in 1991, with Laws 9 and 10 liberalising the electricity industry and facilitating access by independent renewable electricity producers. A subsequent Directive CIP6/92, introduced in 1992, allocated premium buy-back rates for independently generated renewable electricity (paid for via a levy on electricity bills), and resulted in increased interest in grid-connected renewable electricity, especially biomass, wind and PV. Indeed, almost 35% of new capacity authorised in 1992-95 was renewables-based, due to the increased interest shown in renewables by IPPs, and partly due to the priority given to renewable plants by the Laws 9 and 10. Perhaps the most visible manifestation of this policy was seen with the commissioning of Europe's largest PV station (3 MW) at Serre in late. 1994. However, the directive CIP6/92 was abolished in 1997, and new rules for payment of renewable electricity are set by the new Regulatory Authority for Electricity and Gas.

Although efforts to promote renewables emphasise grid-connected electricity production, programmes are also underway to promote service and non-grid electricity applications, especially for PVs. This has proved reasonably successful to date, with Italy having the second largest installed PV capacity in the IEA.

POLICIES

New legal frameworks introduced through Laws 9/91 and 10/91, favourable buy-back rates for independent producers and research, development and demonstration form the three main thrusts of Government support for stimulating the deployment of renewable technologies.

Law 9 partially deregulates independent electricity generation and allows independent power producers to sell renewable electricity to the electricity utility ENEL. Law 10 provides for subsidies of 30-80% of the capital cost of a renewable energy plant. However, financial constraints have meant that few capital subsidies have in fact been disbursed.

Law 10 requires regions to produce regional energy plans and encourages identification of renewable energy potential – including hydro power, energy from waste and district heating (including that based on thermal renewable technologies), and provides incentives for the construction of renewable-based systems for residential, industrial and agricultural purposes. Incentives vary from e.g. 30% for industrial users to 80% of plant cost for isolated housing and lighting. Procedures for approving new cogeneration and renewables capacity were finalised in 1993. Law 10 also includes measures for providing grants for co-generation and renewables-based capacity.

In practice, no subsidies were disbursed under Law 10, and incentives for renewable electricity production were instead provided by the provisions of Law 9, which entered into force via a Directive CIP6/1992. This provided for premium prices to independent producers of renewable electricity for the first eight years of electricity production, and minimum rates guaranteed for the remainder of a project's life. 1997 rates are shown in Table 1. The higher subsidy levels for IPPs (valid for the first eight years of operation only) were designed to reflect the different production costs of different renewable energy systems. The lower level (available from the ninth year of operation) reflect ENEL's avoided cost. The subsidy was financed via a levy on electricity prices. CIP provision 6/1992 also allowed for renewable power to be wheeled within a consortia at a favourable rate (10% below the norm) for 15 years. However, this Directive was abolished in 1997, as part of the government's reorganisation of the electricity sector which aims to cancel all "overprices" on energy tariffs. The new Regulatory Authority for Electricity and Gas has not yet established specific rules regarding future incentives for renewable electricity generation.

The premium prices under Directive 6/1992 led to several proposals for independent, renewable power projects while the Directive was in force. For example, by the end of June 1995, plans for 723 MW of wind capacity had been approved by the Italian Ministry of Industry, Commerce and Trade (MICA), which is responsible for running the government's promotional policies for renewable energy. However, not all of this capacity is likely to be built, due to failure to obtain planning permission. Moreover, much of this capacity is planned to be built in rural, mountainous regions, where the distribution grid would need substantial reinforcement in order to be able to accept significant quantities of wind electricity. Whether or not interest in new renewable energy projects will be sustained under the new regime is yet to be seen.

Since 1993, new independent electricity projects have been subject to index ranking A to E by ENEL which takes account of the overall energy efficiency of the

Table 1
Provisional buy-back rates for electricity (L/kWh)[50] in 1997

Type of plant	Independent electricity producers*	Autoproducers selling surplus electricity only	
		Peak hours**	Off-peak hours***
"A"-rated plant: PV, MSW and biomass	286.1	397.3-440.4	54.7
Wind and geothermal	195.9	247.1-290.2	54.7
Run-of-river hydro <3MW	154.7	178.3-221.4	54.7
Hydro >3MW	290.2 (peak), 54.7 (off-peak)	247.1-290.2	54.7
"B" or "C"-rated plant (or lower): Process fuels and industrial waste	154.7	178.3-221.4	54.7
Oil and gas plants	177.5 (peak), 54.7 (off-peak)	134.4-177.5	54.7
Coal plants	134.6	147.0-190.1	54.7

Source: ENEL

* Price for all plant types after eight years is 98.3 L/kWh.

** Price after eighth year is 84.3 L/kWh.

*** Price after eighth year is 54.7 L/kWh.

plant. Plants using non-waste renewables and other plants with a capacity less than 10 MW are given the highest ranking (A). Waste-to-energy plants are ranked A (for MSW plants) or C (for industrial waste). Additional independent capacity is authorised on the basis of this ranking, and of the 8000 MW total new independent renewables capacity authorised in 1992-95, wind, biomass, waste and hydro projects amounted to 2800 MW, including 723 MW for wind plants.

Despite the increase in premium rates for wind (and other renewable) electricity in 1995, administrative barriers represent a significant hurdle to increased renewables development. Obtaining a construction permit can take up to two years. Using wind as an example, as there are no references to wind installations in local planning regulations, obtaining a construction permit is dependent on the local commune's opinion of the project as well as on obtaining permission from the Air Force and Civil Aviation authorities, the Ministry of Industry and Commerce and the Office of Manufacturing Taxation (UTF). For PV systems, compliance with building codes may result in over-specification of the support structure, with consequent cost increases of the system as a whole.

Italy's first and second National Communications include information on the measures described above, and also recommend that increased information on

---

50 On average in 1997, 1703 L = 1 US $

climate change should be made available to target groups, including the industrial sector and schools/universities. However, no information programmes have been introduced to date.

R&D is the second main thrust for stimulating renewable deployment and penetration, and is concentrated on PV and wind. ENEA and ENEL are the main bodies responsible for implementing R&D policy. In 1996, almost 15% of the total 43.7 M USD Government funds available for R&D were allocated to renewable energy directly, although total funding could be higher as some indirect funding could be available through R&D funds for cross-cutting activities. However, budgetary constraints meant that total energy R&D (including renewables) was 6% lower than in the previous year.

Limited support for liquid biofuels has been available since 1993. A fixed annual quota of diester is exempt from excise tax (the amount was set at 125 ktons of the annual 300 ktons production in 1994). Subsidies of L 15 billion were allocated by the government for the production of diester from set-aside land. Ethanol, however, does not benefit from any tax exemptions, so its production is negligible. Building-integrated PV systems are also encouraged by reducing their VAT rate from 20% to 10%.

In 1995, the Interministerial Committee for Economic Planning approved the *National Plan for Sustainable Development* (aiming to stabilise CO_2 emissions at 1990 levels by 2000). This should maintain policy pressure to increase the penetration of renewable energy. The 1988 *National Energy Plan* had also set out quantified targets for some renewable energy sources, e.g. wind.

STATUS OF RESOURCE EXPLOITATION

Non-hydro renewable energy: largely geothermal and biomass, contributed 2.3% of Italy's total energy supply and 1.9% of total electricity supply in 1996 compared to IEA averages of 3.9% and 2.0% respectively. Geothermal energy is an important renewable energy source in Italy, and is mainly used for electricity production. The majority of biomass is used directly in the residential sector. Small (but rapidly growing) quantities of wind and solar energy are also used in Italy. Hydropower contributed 2.2% of Italy's energy supply as 42 TWh (17.6%) of electricity production in 1996.

Biomass

Italy used 1.2 Mtoe of solid biomass, and small quantities (47 ktoe) of other biomass in 1996. The majority of this was in the residential sector, although significant quantities were also used in coke ovens (107 ktoe) and in industry

(136 ktoe). Some biomass was also used in electricity and CHP plants, and produced 255 GWh electricity in 1996. Reported biomass use has increased substantially over the 1990s, and the importance of biomass-generated electricity is likely to grow, given the amount of additional independent capacity that has been authorised but not yet constructed. A 12 MW biomass gasification plant that will be fired from agro-forestry waste is being constructed near Pisa, as part of an "Energy Farm". In addition, some biogas extraction takes place providing small quantities of electricity and heat.

Wastes

The use of municipal and industrial wastes in Italy varies from year to year, but has contributed approximately 0.1% of total energy supply throughout the 1990s, and produced 416 GWh in 1996.

Wind

Wind-generated electricity is very small in Italy, producing 33 GWh of electricity in 1996, up from 1 GWh in 1993. However, the rate of capacity additions are picking up, with installed capacity growing to 22 MW in 1994, 70 MW in 1996 and a further 71 MW under construction. ENEL-ENEA estimate that total installed wind capacity in 2000 will reach 200 MW, falling significantly short of the 300 MW target, as set out in the 1988 *National Energy Plan.*

Solar

There are over 7 MW of grid-connected PV plants in Italy, and further small plants in use for water desalination. Capacity will grow with the increased use of rooftop PV systems that are being introduced under the 10,000 roof programme, initiated in 1998. Italy has the largest grid-connected PV plant in the world: 3.3 MW at Serre, which can produce approximately 5 GWh/year. Total PV generation varies significantly from year to year and stood at 6 GWh in 1996.

Italian PV capacity is high relative to other IEA countries. However, despite growing PV capacity, the target of 25 MW of installed PV systems by 2000 is unlikely to be met on schedule.

Hydro

Hydropower generated approximately 42 TWh or 17.6% of Italy's electricity in 1996, and accounted for over 90% of renewable electricity generation in the same year. Capacity has climbed slowly from 12.6 GW in 1990 to 13 GW in 1996 and will grow further as the independent hydro capacity that has been authorised is constructed. However, hydropower is a very mature technology in Italy, and the majority of available large-hydro sites have been exploited. The optimistic PEN 2000 target of 23 GW installed hydro capacity is therefore unlikely to be met on schedule.

Table 2
Trends in renewable energy supply and use

	Unit	1990	1992	1995	1996	2000	1990-1996 (%)[1]	1996-2000 (%)[1]
Renewable TPES (excl. hydro)	ktoe	3033	3413	3570	3765	4970	3.7%	7.2%
Percentage of TPES	%	2.0	2.2	2.2	2.3	2.8		
Geothermal	ktoe	2048	2199	2183	2370	3300	2.5%	8.6%
Solar, Wind, Wave, Tide	ktoe	5	6	8	11	200	12.1%	109%
Biomass and Wastes[2]	ktoe	980	1209	1378	1385	1470	5.9%	1.5%
– Biomass	ktoe	763	1074	1222	1210	n.a.	8.0%	n.a.
– Wastes	ktoe	217	135	156	175	n.a.	-3.5%	n.a.
Renewable electricity generation (excl. hydro)	GWh	3504	3719	3900	4472	7430	4.1%	13.5%
Percentage of total generation	%	1.6	1.7	1.6	1.9	2.7		
Geothermal	GWh	3222	3459	3436	3762	3840	2.6%	0.5%
Solar, Wind, Wave, Tide	GWh	6	11	14	39	2210	36.6%	174%
Biomass and Wastes[2]	GWh	276	249	450	671	1380	16.0%	19.8%
– Biomass	GWh	14	18	95	255	n.a.	62.2%	n.a.
– Wastes	GWh	262	231	355	416	n.a.	8.0%	n.a.
Renewable TFC (excl. hydro)	ktoe	699	985	1046	1000	980	6.2%	-0.5%
Percentage of TFC	%	0.6	0.8	0.9	0.8	0.7		
Geothermal	ktoe	0	0	0	0	0	n.a.	n.a.
Solar, Wind, Wave, Tide	ktoe	5	5	7	7	10	7.0%	8.7%
Biomass and Wastes[2]	ktoe	694	980	1039	993	970	6.2%	-0.6%
– Biomass	ktoe	694	980	1039	993	n.a.	6.2%	n.a.
– Wastes	ktoe	0	0	0	0	n.a.	n.a.	n.a.
Hydro TPES	ktoe	2720	3629	3249	3615	4200	4.9%	3.8%
Hydro electricity generation	GWh	31626	42200	37782	42037	48840	4.9%	3.8%
Percent of total generation	%	14.8	18.9	15.9	17.6	17.5		

Notes:

1. Annual Growth Rate

2. Including Animal Products and Gases from Biomass

Geothermal

Geothermal energy is more important to Italy's energy balance than to any other European country, contributing 1.5% of Italy's TPES in 1996, the highest in the IEA after New Zealand. Geothermal energy was used to produce 3762 GWh of electricity and 8916 TJ heat in 1996. Generating capacity has been stable since 1992. However, ENEL-ENEA estimate projected geothermal capacity to increase to 680 MW by 2000 and 700 MW by 2010: significantly under the PEN target for 2000 of 1200 MW capacity, 9 TWh/y output, and 0.33 Mtoe/y from geothermal heat.

Wave and Tidal

No energy use reported.

JAPAN

OVERVIEW OF RENEWABLE ENERGY POLICY

Japan is dependent on imports to satisfy approximately 80% of its energy supply, which makes security of supply an issue of crucial importance. Energy policy emphasises energy conservation and efficiency as well as the continued expansion of nuclear power. Non-hydro renewable energy accounted for 2.0% of Japan's energy supply in 1996 (of which approximately two-thirds is biomass, and one third is geothermal). Increased use of new renewables is an important policy objective which will help to address both energy security and CO_2 concerns, although to a limited extent. Japan has set an ambitious target of increasing the proportion of renewable energy (excluding hydro and geothermal) to 3% of total primary energy supply by 2010. This will require a near trebling of solar thermal heat and significant increases in the use of municipal wastes. In addition, the government has set extremely ambitious targets for the long-term use of individual renewable energy sources, especially for solar energy (both PV and heat) and geothermal.

In order to achieve the renewable energy objective, the national government has put in place several economic incentives for increased renewable energy penetration. These include subsidies, tax incentives and low interest loans for operators of renewable energy systems, as well as financial incentives to local governments that develop or maintain renewable energy systems. In addition, Japan has a guaranteed market and favourable buy-back rates for electricity produced from PV, wind and wastes and has recently facilitated entry of these systems into the power generation sector. Other measures used to promote renewables include the *New Sunshine R&D* programme (which aims to reduce costs of harnessing renewable energy), legislation/guidelines on the environment and actions to reduce non-cost barriers to the increased use of renewables. However, although the use and importance of renewable energy has increased over the last few years, this will have to accelerate rapidly if the short-term targets set by the government are to be met.

POLICIES

Japan's plans to develop renewable energy resources have emerged from an energy policy aiming to maintain economic stability and security, in particular through emphasis on managing and reducing oil import dependency. Continued emphasis on the development of nuclear capacity followed substantial growth in electricity

Table 1
Renewable Energy Targets in Japan

	Target for 2000 (MW)*	Target for 2010 (MW)
Hydropower	45500	57000
Geothermal	600	2800
Solar PV	400	4600
Solar heat	1.1 (Mtoe)	5.5 (Mtoe)
Wind power	20	150
Waste	2000	4000

Source: MITI

* Except for solar heat, which is in Mtoe.

consumption during the 1980s. Recent emphasis on the integration of environmental and energy objectives favours expansion of energy sources with low or no CO_2 emissions, in particular nuclear but also renewables. The *Action Plan to Arrest Global Warming* outlines the broad lines of support available to renewable energies, and the *Law concerning Special Measures to Promote the Use of New Energies* was enacted in June 1997.

Promotion of renewable energy is seen as indispensable in order to meet Japan's aims of returning per capita CO_2 emissions to their 1990 level by 2000. Japan has set up several ambitious, but non-binding, targets for increased renewable energy penetration (Table 1), and these form part of Japan's official energy outlook.

The New Energy and Industrial Technology Development Organisation (NEDO) is responsible for implementing renewable energy policy via the *New Sunshine Project*. This project aims to work towards sustainable growth by developing and reducing the costs of innovative technology. However, the ultimate responsibility rests on the Ministry of International Trade and Industry (MITI). The New Energy Foundation (NEF) also assists in renewable policy implementation. The total R&D budget for the *New Sunshine Programme* is estimated at 56 billion JP¥ in 1996[51].

The legislation enacted in 1997 specifically to promote renewables sets the framework for government policies to promote renewable energy, and outlines the measures to be undertaken by the suppliers and users of energy (e.g. an effort to buy renewable electricity). It also formulates new guidelines for renewable energy uses.

Policy for renewable energy promotion rests on three pillars: a range of financial incentives from national government (in the form of subsidies, tax exemptions, and

---

51 In 1997, 1 US $ = JP¥ 121

Table 2
Legislation and Programmes to encourage renewable energy in Japan

	Implementation date	Description
Basic Environment Law	Nov. 1993	Stipulates basic principles about environmental conservation and lays out the responsibilities of the State, local governments, corporations and the people. The Law prescribes setting up a Basic *Environment Plan,* which was done in December 1994. This sets four long-term objectives, including a "Sound Material Cycle" to minimise waste.
Basic Guideline for New Energy Introduction	Dec. 1994	Guidelines to promote the introduction of renewable energy were determined by the Council of Ministers.
The New Sunshine Programme	1993	Encourages energy and environmental technology development and deployment.
Amendment of the Electric Utility Law	1995	Facilitates entry into the power generation sector and reforms fee regulations.
Law on Special Measures to Promote Use of New Energies	1997	Outlines measures for energy suppliers, users and renewable energy manufacturers to promote accelerated introduction of new energies.
The Packaging Recycling Law	1995	Encourages consumers to separate wastes and business to recycle the separated waste streams.
The Action Plan for Greening Government Operations	1995	This 5 year plan aims to increase energy efficiency and conservation and reduce wastes. The only reference to renewable energy is "efficient use of solar energy" in the construction of new buildings.
Action Program to Arrest Global Warming	1990	This programme aims to stabilise per capita CO_2 emissions at 1990 levels by 2000, and outlines measures that should be taken to achieve this goal such as the promotion of renewable electricity, development of renewable technologies and increased energy efficiency.

low-interest loans) for renewable energy installations and R&D assistance from government. Actions to reduce non-cost barriers, such as planning consent, have also been taken by the government. A number of laws and plans that impact renewable energy are set out in the table above. These are not discussed in detail below, because the legislation tends to set out principles for future policy directions, rather than outline specific incentives for renewables. MITI also issued a request to several industrial associations in 1992 to develop voluntary plans in order to incorporate environmental criteria – including increased use of renewables – into their decision making process. By October 1994, 362 companies had drawn up such a plan.

The financial incentives available to promote renewable energy can be very generous – especially for demonstration plants. The level of incentive varies on the renewable energy source being promoted, but ranges from 10% for hydro systems to 50% for some solar PV and wind installations. Incentives for hydropower are shown in Table 3, and those for other renewable energy sources are outlined below. For hydro development, consent is needed from the national government for the use of water from large rivers and from local government for small rivers is needed.

Significant financial support for PV systems is available to individuals and for public buildings. From FY 1994, the government has subsidised up to half the installation cost of a small (3-4 kW) PV system in individual dwellings. This subsidy was capped at 0.34 M JP¥ in 1997 (and has been decreased over the years). As a result,

Table 3
Financial Incentives in place for small and medium hydro plants

Size of plant	Incentive	Comment
Subsidies		
<5000 kW	20% of capital cost	This was increased from FY
5000 kW < 30000 kW	10% of capital cost	1994 to 30% and 20% respectively for less economically attractive sites.
Reimbursement of a proportion of the interest payments: < 5000 kW	3% of total interest payments in years 1-3, 2.5% in years 4-6, 1.5% in years 7-9 and 1% in years 10-12.	This subsidy is paid by the New Energy Foundation (a private body but partially funded by government).
5000 kW < 30000 kW	2% of total interest payments in years 1-3, 1% in years 4-6, 1% in years 7-9.	Less economically attractive sites benefit from increased subsidies (up to a maximum of 15 years, and at higher levels).
All small and medium hydro	50% of capital cost	This incentive applies to new hydro technologies only.
Tax incentives (all small and medium hydro)	The developer can opt either for a reduction in income tax up to an amount equalling 7% of the plant's capital cost, or for a depreciation allowance of 30%.	
Low Interest Loans (all hydro < 100 MW)	Loans of 2.4% (up to 40% of the total cost of the plant).	These loans are obtained from the Japanese Development Bank.
Financial incentives for local government (plants > 1MW)	Grants of 250 JP¥ /kW for new construction and 0.075 JP¥/kWh for facilities already in place	Disbursed from national to local government.

subsidies of over 9.4 billion JP¥ were disbursed in 1994-6 (corresponding to the installation of 3500 such systems) and a further 9400 systems were expected to be installed at a cost of 11.1 billion JP¥ in 1997. Public buildings have benefitted from subsidies of up to 67% for PV systems from 1992. However, the uptake has been lower than for private dwellings.

In order to attain the 150 MW capacity target for wind by 2010, the government has set up a National Test Project, run by NEDO, with associated subsidies. For wind systems, these are:

■ 100% subsidy for meteorological measurements (for selected projects);

■ 50% capital subsidy for the design of a pilot demonstration plant; and

■ 50% capital subsidy for the construction of this plant.

There are three similar subsidies available for geothermal system development, although that for construction of the plant is 20% (rather than 50% for wind). Support is also available for feasibility studies on the development of distributed energy sources including projects which use waste heat and renewable energy. This is limited to 10 M JP¥ per study. Prototype systems benefit from a capital subsidy of 15% of the capital cost (up to a maximum of 600 M JP¥).

In addition to the financial incentives outlined above, Japan has encouraged electricity produced from PV, wind, waste and fuel (although there is no guaranteed market). Buy-back rates for PV and wind electricity are equal to consumer end-prices, while that for waste power is at a premium rate.

The government of Japan also want to encourage the use of clean energy vehicles, which include vehicles that run on biofuels as well as on electricity, gas or LPG. However, the focus of this policy is to reduce transport oil demand rather than to specifically encourage transport biofuels.

The government promotes penetration and utilisation of new energy technologies through the New Sunshine R&D Programme which began in 1993. However, renewables R&D, which totalled 13.2 billion JP¥ in 1997, only accounts for 3% of total energy-related R&D (the vast majority – over three-quarters – is spent on nuclear power research). Of government renewable R&D funding, solar electricity research receives most support (an estimated 8.2 billion JP¥ in 1997), followed by geothermal energy. The solar energy programmes aim to reduce the production cost of PV to 100-200 JP¥/Wp by 2000, compared to 600 JP¥/Wp in 1992. In addition to research on domestic uses of PV, NEDO undertakes joint research on practical uses of PV power generation systems in Nepal, Mongolia, Thailand and Malaysia.

Certain measures to reduce non-cost barriers are in place. A reporting requirement was eased in 1995 by waiving the requirement for autoproducers whose generation is less than 1 MW. Efforts have also been made to simplify the requirements concerning safety and technical inspections, in some cases allowing organisations to regulate these aspects autonomously. The Agency of Natural Resources and Energy is promoting the introduction of dispersed power sources, with guidelines in place since 1993 on the supply of excess power via the interconnection of small scale power systems from companies and individuals.

Efforts to promote public awareness through education, dissemination of information and cooperation between the government (via NEDO) and various publicity agencies have focussed on the environment in general as well as energy conservation and global warming.

No evaluation of the policies described above to promote renewable energy has been undertaken to date. However, the government estimates that the most important barrier to enhanced renewable deployment is financial cost.

STATUS OF RESOURCE EXPLOITATION

The contribution of non-hydro renewable sources to total energy supply was 10.4 Mtoe or 2.0% in 1996, a slight increase in importance from 1990. The majority of this was from the use of biomass and wastes, although a significant amount of geothermal energy (3.4 Mtoe) was also harnessed. Most renewable energy is used for electricity generation, and this emphasis is likely to remain despite continued direct use of biomass, and increased use of solar thermal (heat) and geothermal heat in the residential sectors. In addition, hydropower generated 8% of Japan's electricity and contributed 1.4% to total energy supply in 1996.

If Japan's ambitious renewable energy targets are to be met, there will also have to be a huge expansion of renewable electricity. The government expects that the majority of this will come from more autoproducers (including greatly increased distributed electricity generation, e.g. from rooftop PV systems). Moreover, a huge push will be needed before the turn of the century if these renewable energy targets are to be met. This is especially true for solar PV, where the target is over ten times the reported capacity in 1996.

Biomass

Use of biomass for energy purposes has been approximately stable since 1990, at around 6.4 Mtoe. Around half of this is used directly, mainly in industry, and the remainder is used for electricity generation. Electricty production from biomass has been growing steadily since 1990, and needs to continue growing fast if the government's target for electricity capacity (which is for the combined total from biomass and wastes) is to be met.

Waste

The use of municipal waste for energy purposes has been growing steadily and reached 652 ktoe (to generate approximately 3.7 TWh) in 1996. As for biomass, this should increase sharply in the near future if government targets for biomass and wastes (aggregated) are to be met.

Wind

Despite Japan's wind potential spanning the country with strong winter winds in the North and typhoons in the Southern islands, areas where wind power can be introduced are limited. Total installed capacity is small, but has been growing steadily since 1990 and, was quoted in the *IEA Wind Energy Annual Report 1996* as standing at 13.9 MW by the end of 1996 (although only 1 MW generating 2 GWh was reported officially to the IEA). This capacity is made up of mainly small turbines (under 400 kW). However, NEDO has also constructed a number of larger demonstration plants including one of 1.7 MW. The wind resource has been mapped since 1983, and a wind atlas was produced in 1995.

Solar

Government projections assume that Japan meets its extremely ambitious PV target of 400 MW by 2000 and 4.6 GW by 2010. Although no PV capacity was reported to the IEA, NEDO estimates that total installed capacity was approximately 38 MW at the end of 1995. Moreover, the number of rooftop PV systems installed between 1994-1996 correspond to approximately 13 MW capacity, and if plans for 1997 were achieved, capacity of rooftop systems alone could have reached 30 MW by the end of that year. However, even if this was the case, Japan is still a long way from its near-term target, and it looks unlikely that this will be met on time.

Hydro

Hydropower, excluding pumped storage, generated 80.5 TWh (8.0% of total electricity) in 1996. Both capacity and generation are expected to expand significantly in the long-term: capacity is expected to increase by over 4GW from its 1996 value to 25 GW by 2010, and generation (excluding from pumped storage) is projected to grow to 105 TWh in the same year.

Geothermal

Japan is one of the IEA countries that possesses significant geothermal resources. The majority of the 0.53 GW installed capacity in 1996 is public sector owned. Only a further 70 MW of utility capacity are expected to be commissioned by 2006, and, unlike other renewable energy sources, it is utilities that are expected to develop the majority of geothermal capacity. However, geothermal energy use has increased significantly (in line with capacity increases) since 1994 and reached 3.7 TWh in 1996.

Table 4

Trends in renewable energy supply and use

	Unit	1990	1992	1995	1996	2000	1990-1996 (%)[1]	1996-2000 (%)[1]
Renewable TPES (excl. hydro)	**ktoe**	**8214**	**8220**	**9885**	**10449**	**11762**	**4.1%**	**3.0%**
Percentage of TPES	%	1.9	1.8	2.0	2.0	2.4		
Geothermal	ktoe	1497	1537	2916	3398	3520	14.6%	0.9%
Solar, Wind, Wave, Tide	ktoe	0	0	0	0	43	n.a.	n.a.
Biomass and Wastes[2]	ktoe	6717	6683	6969	7051	8199	0.8%	3.8%
- Biomass	ktoe	6368	6270	6431	6399	n.a.	0.1%	n.a.
- Wastes	ktoe	349	413	538	652	n.a.	11.0%	n.a.
Renewable electricity generation (excl. hydro)	**GWh**	**12904**	**13846**	**22911**	**23949**	**n.a.**	**10.9%**	**n.a.**
Percentage of total generation	%	1.5	1.6	2.3	2.4	n.a.		
Geothermal	GWh	1741	1787	3173	3673	4035	13.2%	2.4%
Solar, Wind, Wave, Tide	GWh	1	1	1	2	500	12.2%	298%
Biomass and Wastes[2]	GWh	11162	12058	19737	20274	n.a.	10.5%	n.a.
- Biomass	GWh	9106	9695	16677	16602	n.a.	10.5%	n.a.
- Wastes	GWh	2056	2363	3060	3672	n.a.	10.1%	n.a.
Renewable TFC (excl. hydro)	**ktoe**	**3700**	**3512**	**3636**	**3658**	**6018**	**-0.2%**	**13.3%**
Percentage of TFC	%	1.3	1.1	1.1	1.1	1.7		
Geothermal	ktoe	0	0	187	239	50	n.a.	-32%
Solar, Wind, Wave, Tide	ktoe	0	0	0	0	0	n.a.	n.a.
Biomass and Wastes[2]	ktoe	3700	3512	3449	3419	5968	-1.3%	14.9%
- Biomass	ktoe	3700	3512	3449	3419	5968	-1.3%	14.9%
- Wastes	ktoe	0	0	0	0	0	n.a.	n.a.
Hydro TPES	ktoe	7680	7099	7062	6925	7391	-1.7%	1.6%
Hydro electricity generation	GWh	89305	82545	82118	80522	85942	-1.7%	1.6%
Percent of total generation	%	10.5	9.3	8.4	8.0	8.3		

Notes:

1. Annual Growth Rate

2. Including Animal Products and Gases from Biomass

LUXEMBOURG

OVERVIEW OF RENEWABLE ENERGY POLICY

The importance of non-hydro renewable energy in Luxembourg is 1%: amongst the lowest of any IEA country. Moreover the potential of increased renewable energy use is limited. Nevertheless, it is being encouraged actively via favourable and guaranteed electricity markets, tax deductions, and information campaigns. One of the most important drivers behind increased use of renewables in Luxembourg is their ability to reduce domestic emissions of CO_2.

The measures put in place to promote renewable energies have succeeded in increasing the use of some of these energy sources to a limited extent.

POLICIES

Luxembourg is almost entirely dependent on energy imports and its energy policy pursues energy security through diversification, energy efficiency and environmental protection. Luxembourg has only limited renewable energy potential, as it is a small and landlocked country. Nevertheless, renewable energy is being encouraged by a number of measures including guaranteed electricity markets at favourable rates, and grants for renewable energy technologies.

Key policy developments which relate to renewables include the *Energy Efficiency Law* adopted in 1993 which in turn led to a 1994 Grand Ducal regulation on purchase price regulations for surplus electricity produced by CHP and renewable energy sources. A subsequent ministerial regulation in 1994 established a government programme for new CHP projects and renewable energy sources. Since December 1996, expenditure on certain technologies, including renewable energy systems, is tax deductible. Information and education measures relating to renewables are not followed. No public funding is allocated to renewable (or other) energy R&D.

Renewable energy policy is overseen by the government's Energy Agency, which was created in 1991. The Energy Agency, set up as a company under private law whose shareholders are the Government (50%), CEGEDEL (40%) and SEO (10%), is focusing its renewable energy activities on the modernisation of 11 small hydropower plants in Luxembourg, the use of thermal solar energy in swimming pools, sports installations and camping sites, and the evaluation of the potential for wind power use in Luxembourg. A wind map was completed in January 1994 and two photovoltaic stations of 3 kW each were commissioned in September 1993 to supply electric vehicles.

In relation to capacity expansions to cover marginal electricity consumption growth in the public network, the Government emphasises the promotion of combined heat and power followed, to a lesser extent, by renewables. The policy context for renewables benefits from a number of initiatives undertaken primarily to encourage CHP.

For installations based on wind or active solar energy, biomass, biogas and heat pumps, the December 1994 ministerial regulation allocates various grants according to the capacity of the installation or the end-user. Available capital subsidies are 25% (capped at 60 000 LF[52] per installation) for wind turbines under 50 kW and residential uses of other renewable energies such as solar water heaters and heat pumps. Non-residential uses of solar, biomass and heat pumps also benefit from a 25% subsidy (capped at 1.5 M LF). For wind turbines >50 kW, grid-connection of capacity up to a national total of 5 MW will be subsidised at 3000 LF per installed kW (capped at 6 M LF). By the end of 1997, 17.5 M LF had been disbursed, mainly for solar hot water heaters and wind turbines.

The Grand Ducal regulation of 30 May 1994 sets the remuneration system for electricity produced from CHP and renewable energy sources. The price paid for surplus electricity produced by renewable installations with a capacity up to 500 kW averages 2.95 LF/kWh; for installations from 501 kW to 1 500 kW, the price paid is 2.30 LF/kWh for day supplies and 1.20 LF/kWh for night supplies. There is also a possible annual premium of 4 500 LF/kW installed used for peak power, on the condition of electricity deliveries during the network's peak load. This system corresponds to a bonus of about 20%. In addition, surplus electricity based on wind energy and photovoltaic receives an extra bonus of 1 LF/kWh.

STATUS OF RESOURCE EXPLOITATION

Reported non-hydro renewable energy use amounted to 1 ktoe in 1996 or 1% of TPES, compared to an IEA average of 3.9%. Wood and wastes, mainly used for electricity generation, provide the vast majority of Luxembourg's renewable energy supply, although small amounts of solar heat are also used. In addition, small-scale hydropower generated 60 GWh in 1996 (12% of domestic electricity generation). In 1996, municipal waste accounted for 46% of total renewable TPES and hydropower 13%, providing 39% and 55% of renewable electricity generation respectively.

---

52   On average in 1997, I US $ = 35.76 LF

Biomass

Small amounts (15 ktoe) of wood are reported as being used in the residential sector, and 1 GWh was also generated in 1996 from biogas. This will increase as 4 new biogas installations are under construction.

Waste

Municipal waste was used to produce 42 GWh of electricity in 1996. This represented just over 8% of total domestic electricity generation in the same year.

Wind

Work has been carried out on the potential for wind power in Luxembourg, and although no generation was reported to 1996, the government estimate that 2.7 GWh were generated from 2 MW in 1997. The government estimate wind potential at 1% of electricity consumption.

Hydro

There are no large hydro resources in Luxembourg, and the limited resources on the Moselle river have largely been exploited. Small hydro generation varies widely year by year, and stood at 60 GWh in 1996. Around 5-10% was produced from installed hydro autoproduction.

Solar

Small amounts of PV electricity (6 GWh) were generated in 1995 and 1996. The importance of solar hot water heating should increase as 162 subsidies were allocated to such systems by the end of 1997.

Geothermal, Ocean/Tidal/Wave/Heat pumps

None reported to IEA.

Table 1
Trends in renewable energy supply and use

	Unit	1990	1992	1995	1996	2000	1990-1996 (%)[1]	1996-2000 (%)[1]
Renewable TPES (excl. hydro)	ktoe	25	42	40	34	20	5.5%	-12%
Percentage of TPES	%	0.7	1.1	1.2	1.0	0.6		
Geothermal	ktoe	0	0	0	0	0	n.a.	n.a.
Solar, Wind, Wave, Tide	ktoe	0	0	1	1	0	n.a.	n.a.
Biomass and Wastes[2]	ktoe	25	42	39	34	20	5.2%	-12.1%
- Biomass	ktoe	0	16	16	16	n.a.	n.a.	n.a.
- Wastes	ktoe	25	26	23	18	n.a.	-5.3%	n.a.
Renewable electricity generation (excl. hydro)	GWh	34	33	60	49	40	6.3%	-4.9%
Percentage of total generation	%	5.4	5.0	12.1	9.8	2.0		
Geothermal	GWh	0	0	0	0	0	n.a.	n.a.
Solar, Wind, Wave, Tide	GWh	0	0	6	6	0	n.a.	n.a.
Biomass and Wastes[2]	GWh	34	33	54	43	40	4.0%	-1.8%
- Biomass	GWh	0	1	1	1	0	n.a.	n.a.
- Wastes	GWh	34	32	53	42	40	3.6%	-1.2%
Renewable TFC (excl. hydro)	ktoe	0	15	15	15	n.a.	n.a.	n.a.
Percentage of TFC	%	0.0	0.5	0.5	0.5	n.a.		
Geothermal	ktoe	0	0	0	0	n.a.	n.a.	n.a.
Solar, Wind, Wave, Tide	ktoe	0	0	0	0	n.a.	n.a.	n.a.
Biomass and Wastes[2]	ktoe	0	15	15	15	n.a.	n.a.	n.a.
- Biomass	ktoe	0	15	15	15	n.a.	n.a.	n.a.
- Wastes	ktoe	0	0	0	0	n.a.	n.a.	n.a.
Hydro TPES	ktoe	6	6	7	5	10	-2.5%	18.0%
Hydro electricity generation	GWh	70	70	84	60	120	-2.5%	18.9%
Percent of total generation	%	11.2	10.6	17.0	12.0	6.1		

Notes:

1. Annual Growth Rate

2. Including Animal Products and Gases from Biomass

THE NETHERLANDS

OVERVIEW OF RENEWABLE ENERGY POLICY

The Netherlands currently uses very little non-hydro renewable energy: 0.7 Mtoe or 1% of its TPES was met by these sources in 1996. Although this is higher than the value in 1990 (0.4 Mtoe or 0.6% of total energy supply) it is one of the lowest proportions amongst IEA countries, due partly to the lack of space for biomass (energy crop) development, and limited space for on-shore wind turbines. In addition, hydro resources are extremely limited.

The Netherlands' greenhouse gas emission reduction objective is -6% in 2008-2012 compared to 1990 (under the EU burdensharing agreement), and it plans to use increased renewable energy use as one of the means of achieving this objective. The main targets for renewable energy use are for 3% of energy to be renewable-based by 2000 and for a 10% share of renewables in primary energy supply by 2020. The majority of renewable energy policies are therefore directed at harnessing more abundant renewable energies, mainly electricity production from wind, and electricity and heat production from biomass and wastes.

As well as the voluntary agreements that have been widely used between the government and the electricity sector until recently, the 1998 *Electricity Act* has introduced a new policy type to promote renewable electricity: from 2001, consumers may be required to purchase "green certificates" confirming that a certain percentage of electricity consumed is from renewable sources. A wide range of other policy measures to promote renewables are also used, such as economic and fiscal incentives (largely tax exemptions, as capital subsidies have been phased out for some renewable energy applications), guaranteed buy-back rates for renewable electricity, and information dissemination activities.

POLICIES

Dutch energy policy aims to provide reliable, affordable, and clean energy. Almost half Dutch energy is supplied by indigenous natural gas. The densely-populated Netherlands has a shortage of space to exploit diffuse renewable energy sources such as biomass, and a lack of suitable sites for large-scale hydropower developments, which means that the potential for renewable energy development is limited. However, available resources of wind, waste and biomass mean that continued growth in renewable energy use is feasible: the short-term objective is to produce 3% of energy from renewables by 2000. The longer-term target is for renewables to supply 10% of energy by 2020: this would be mainly met from renewable-generated electricity, but to a lesser extent the use of solar water heating, ambient heat and geothermal energy. Alternative transport fuels are not expected to become economically attractive.

Renewable energy policy has evolved rapidly in the Netherlands over the 1990s. The current policy framework to encourage renewable energy is laid out in the 1997 White Paper *Renewable Energy – Advancing Power*. Although this action programme includes voluntary agreements with the electricity distribution companies and industry, other measures (such as tax exemptions and other fiscal instruments, increased funding for renewables R&D) aiming to increase the market penetration of renewable energy are also included. The White Paper also laid the foundations to amend the 1989 *Electricity Act*. The revised 1998 *Electricity Act* includes an option for requiring electricity distributors to distribute a certain amount of "sustainable" electricity at a future date. In addition, seven electricity producers offer the possiblity of buying "green power".

The Ministry of Economic Affairs, EZ, is responsible for promoting renewable sources of energy. Voluntary agreements with utilities are an important component, but may become less so with the introduction of "green certificates" (above). Economic and fiscal incentives, regulations, information and R&D are also used widely. The Ministry of Housing, Physical Planning and Environmental Management, VROM, is also involved in some renewable energy policy issues, and works with EZ on planning and siting problems for wind turbines and the joint energy-from-biomass programme. Recent policies and legislation that impact renewable energy development are outlined below.

Renewable Energy – Advancing Power outlines the measures needed to increase renewable energy supply in the Netherlands, and restates the expected supply of each renewable energy source to 2020 (although interim targets are often lower). This White Paper is predominantly a "technology push" programme that aims to improve the price-performance ratio, to promote market penetration of renewables, and to reduce administrative bottlenecks. While the programme includes voluntary actions, such as agreements between government and utilities, it also sets out a number of different economic and fiscal incentives used in the promotion of renewable energy. For example, there are many types of tax exemptions including corporation tax exoneration for investments in renewable energy technology and income tax exoneration for investments in "green" investment funds. The programme also doubled renewable R&D funding. However, the capital subsidies that were available for wind turbines and heat pumps during the 1990s were phased out in 1996, although they are still in place for solar hot water heaters. The manner in which renewable energy is promoted is therefore changing from a system emphasising voluntary agreements and capital subsidies, to one that is more market-based and that gives a more prominent role to fiscal incentives.

Table 1

Summary of recent policies relevant to renewable energy promotion in the Netherlands

The Dutch Electricity Act, 1998	Includes provisions for "green certificates" for renewable electricity and the possibility to set an obligation for consumers to buy an amount of renewable electricity.
Renewable Energy-Advancing Power: Action Programme 1997-2000, 1997	Government white paper on renewable energy
CO_2 reduction plan, 1996	1500 M NLG, of which 1000 M NLG for CO_2-limiting projects and 500 M NLG for Joint Implementation, energy conservation and R&D.
Regulatory energy tax, 1996	Renewable energy is exempt from this tax, which has significantly increased the costs of small-scale energy consumption from other sources.
Environmental Action Plan, MAP 2000, 1997 (follow-on from 1994 plan)	This agreement between the government and the energy distributors lists actions the latter will take to reduce CO_2 emissions.
Energy Investment Relief, 1997	Tax relief in renewable energy technology and energy conservation technologies. Investments may be offset against taxable profits at a rate varying between 40-52%.
Long term agreement for solar water heater market development (1994-1997)	Agreement between all relevant market parties to install 80,000 solar water heaters by 2000. This agreement is in the process of being updated.
Agreement on Solar PV, 1997-2000	Agreement between all relevant market parties to have 7.7 MWp (77000 m^2) installed in 2000.
"Green electricity"	A regional, voluntary policy whereby consumers can opt to pay more for electricity produced from renewables.
Renewable Energy Project Bureau	Project bureau for the promotion of and information about renewable energy.

The 1998 *Electricity Act* contains provisions that explicitly aim to promote the amount of "sustainable electricity", defined as electricity from mini (<15 MW) hydro, wind, solar or biomass sources, and are classified as "green" by the government. Currently, individual distribution companies have undertaken commitments to either buy or produce 1.7 TWh of sustainable electricity ("green labels") by 2000. This compares to the 1996 level of under 0.8 TWh. This will result in a developing market for renewable electricity as producers work towards meeting their commitments.

From 2001 there may be a statutory obligation on all consumers to buy a certain number of "green certificates" (i.e. corresponding to a certain amount of green power). In other words, implementation of the *Electricity Act* will effectively require a certain amount of the electricity supplied to customers in the Netherlands being produced from sustainable energy sources. This percentage is decided by the government, and green certificates are awarded to renewable electricity generators who are able to sell them. This creates a market for these certificates, and enables renewable electricity producers to sell to consumers in any part of the country, or to an exchange. In 1998, the Dutch government's target for the contribution of renewable power to the total Dutch supplies is 3.2% in 2003, rising to 17% in 2020. Foreign producers will be also allowed to apply for green certificates, although, as for Dutch producers, they will have to contract with suppliers or consumers within the Netherlands so that they are not elegible for "green" incentives where they are situated as well as in the Netherlands.

The *Third White Paper on Energy Policy* set targets for individual renewable energy contributions to Dutch TPES for the years 2000, 2007 and 2020, and outlined the policies that will be undertaken to achieve these targets. The Paper's renewable energy targets for 2000 are unchanged from those established in 1993 (in the *Second Memorandum on Energy Conservation*, SMEC), although longer-term targets have been brought forward (in general, targets for 2010 laid out in the SMEC now apply to 2007) and, in some cases, strengthened. Only the target for geothermal energy has been lowered.

The energy distribution sector's *Environmental Action Plan, MAP,* was produced in consultation with central Government and includes information on measures that will be undertaken to stimulate the uptake of renewable electricity from wind power, hydro, biomass, PV, solar thermal power and waste incineration. These actions aim to reduce emissions by 2.7 Mt CO_2 by the year 2000, by producing 3.1% of electricity sales and 0.1% of gas sales from renewable sources. The additional costs of increasing renewable energy use will be paid for from the "MAP supplement", contributed to by the distribution companies and the Government. The distribution companies (EnergieNed) will fund their contribution from an environmental surcharge (capped at 2.5%) levied on their customers.

In 1995 and 1996, MAP funds also helped pay a subsidy on wind energy from privately owned turbines. Total payment amounted to 0.163 NLG/kWh over a 10 year period for new projects up to 2 MW that had not benefitted from other government subsidies and up to 0.133 NLG/kWh for projects < 2 MW that had benefitted from government subsidies. Buy-back rates for larger projects were negotiated on a case by case basis, composed of a minimum of 0.109 NLG/kWh and a contribution from the MAP funds. From 1998, buy-back rates from 1998 were agreed with the utilities. Buy-back rates for PV electricity varies, but "net metering" (allowing a consumer's meter to run backwards as compensation for feeding electricity into the grid) is practised by some distribution companies. Buy-back rates will be phased out for some systems with the introduction of green certificates, although wind and solar producers with a capacity under 8 MW at end 2001 or more than 600 kW in 2006 will continue to benefit from guaranteed buy-back rates.

Waste policy is the responsibility of VROM, and the priority is to minimise waste formation by reusing and recycling as much as possible. Landfilling of combustible waste was banned in 1997. Waste policy is important as regards renewable energy potential as incineration of waste produces the majority of the Netherlands' renewable energy (either via electricity production or from heat production). The *Energy Recovery from Waste and Biomass* programme aims to improve the gasification, combustion and digestion of biomass and thereby facilitate its market introduction.

The policies by which renewable energy in the Netherlands are promoted has shifted significantly since the early 1990s. Capital subsidies were phased out for wind turbines at the end of 1995 and for heat pumps end 1996. The subsidies for solar water heaters will be continued until 2001. Current promotional measures emphasise competition within renewable energy supply, and favourable tax treatment for renewable energy investments and expenditure.

An innovative policy encouraging renewable electricity, since replicated in many other European countries, was initiated by a regional electricity company, PNEM, and is now used by the seven largest distribution companies in the Netherlands. Research indicated that a significant proportion of customers would be prepared to pay a small premium for *"green electricity"*. PNEM have therefore set up a scheme whereby customers pay a price differential of 0.089 NLG/kWh to receive power from renewable energy sources (this amounts to approximately 22 NLG/month for an average family). Revenue from this policy is used to increase renewable generating capacity in the area. The rate of VAT on Green Electricity is charged at 6% instead of the normal 17.5%.

The *Administrative Agreement on Wind Energy Siting* aims to ease siting and planning problems for wind projects. In order to do this, the Government and the seven provinces with the highest wind potential have agreed on capacity targets for 2000 by province. A public education campaign via the National Bureau of Wind Energy has been initiated, as part of the 1997 Project Bureau for Renewable Energy, to increase support for wind-generated electricity. Within the scope of the CO_2 reduction plan, a demonstration project for the installation of 100 MW offshore windpower was launched in 1997. The Ministry of Economic Affairs has invested 60 M NLG in this project.

The Dutch see R&D as one of the few areas where financial instruments such as subsidies remain necessary to encourage the deployment of technologies that are in the demonstration phase. The Dutch also place importance on international co-operation in research, and work with the EU and the IEA (via Implementing Agreements) in a number of areas. One of the more active domestic policies is NOVEM's Dutch solar energy research programme (NOZ) that aims to reduce the price and increase the efficiency of solar cells. The level of renewable R&D has fluctuated significantly over the last few years, being cut from 105 M NLG in 1993 to 55 M NLG in 1995, and then increased to 110 M NLG in 1998.

The Dutch Government set up an investment scheme at the beginning of 1995 aimed at increasing funds available for nature and environment-related projects (including renewable energy) in the Netherlands. Seven banks have responded, setting up *"Green Funds"*, 50-60% of interest and dividends from which is exempt from income tax. The projects which are considered for green funding have to be recognised as "green" by the Ministry of the Environment.

STATUS OF RESOURCE EXPLOITATION

The importance of non-hydro renewable energy has increased slowly since 1990 to reach 1% of of TPES in 1996. The majority of this comes from the utilisation of municipal waste and landfill gas to generate heat and electricity, although the contribution by wind power is increasing.

Biomass

Supply and consumption of biomass in the Netherlands is limited. Wood use was estimated at 168 ktoe in 1996, all of which is used directly, mainly in the residential sector. Sludge/sewage gas is also used for electricity generation (270 GWh), and heat generation of 210 TJ. The use of solid biomass (i.e. wood) for electricity generation is also likely to increase.

Waste

Municipal waste is the most important renewable energy source for the Netherlands. Production of electricity from these sources was 1.8 TWh (2.1% of total electricity generation) in 1996, a significant jump from its value in previous years. Waste is a significant source of Dutch renewable energy. Dutch policy aims at significantly increasing the use of waste for energy purposes to 1075 ktoe by 2020.

Wind

The ambitious 750 MW windpower target for 2000 looks unlikely to be met on time unless installation rates grow rapidly: installed capacity was only 333 MW in mid-1998. This is also below the level of the target for 1995 (400 MW of wind capacity). The 1997 *Renewable Energy Action Programme* indicates expected wind capacity to be 750 MW in 2000 and 2000 MW by 2007. Although the cost of wind power has fallen from NLG 0.7/kWh in 1980 to 0.25/kWh in 1990 and

0.18/kWh in 1995, the market for wind turbines needs to increase before economies of scale drive costs down further. In addition, finding suitable sites for wind turbines has been a problem.

Solar

The installed capacity of PV systems in the Netherlands grew to 3 MW by the end of 1996. The majority of these systems are stand-alone, with only 0.3 MW grid-connected (including those installed under the *"10-roofs project"* at Heerhugowaard. Further (larger) demonstration projects of PV in buildings were initated in 1996, e.g. the 1 MWp programme in Amersfoort. However, the Dutch government have ambitious targets for increased capacity of PV: 7.7 MW by 2000, rising to 100 MW by 2007 and 1.4 GW in 2020. The 2020 target is equivalent to 560,000 households being equipped with a 20 m² rooftop PV system.

Hydro

The Netherlands' hydro potential is extremely limited due to its topography: the Dutch Electricity Generating Board (SEP) estimate the technical potential at 100 MW, although the economic potential has been estimated at 53 MW. Nevertheless, hydro generation has increased from 3 GWh in 1985 to an estimated 80 GWh in 1996, peaking at 120 GWh in 1990 – a figure that is projected to increase to 210 GWh by 2010.

Geothermal

The Netherlands possesses economically-viable potential for electricity generation from geothermal energy, but this has not been developed at present. A demonstration geothermal project that will produce hot water for heating greenhouses has been proposed by energy distribution companies.

Passive solar, heatpumps

The use of ambient heat via heatpumps is forecast to grow from its 1994 level of almost 17 ktoe to 170 ktoe by 2000 and 2400 ktoe by 2010. This growth is expected to arise from the *Dutch Heat Pump Action Programme*, launched by Novem in the framework of the SMEC. The programme should lead to 24,000 heat pumps for space heating of individual houses, 20 collective space heating systems for domestic houses and hot water supply by 100 000 heat pump boilers by the year 2000.

Use of passive solar is encouraged indirectly via Energy Performance Norms for buildings.

Table 2
Trends in renewable energy supply and use

	Unit	1990	1992	1995	1996	2000	1990-1996 (%)[1]	1996-2000 (%)[1]
Renewable TPES (excl. hydro)	**ktoe**	**382**	**539**	**596**	**730**	**2000**	**11.4%**	**28.7%**
Percentage of TPES	%	0.6	0.8	0.8	1.0	2.6		
Geothermal	ktoe	0	0	0	0	0	n.a.	n.a.
Solar, Wind, Wave, Tide	ktoe	6	15	31	46	100	40.4%	21.1%
Biomass and Wastes[2]	ktoe	376	524	564	683	1900	10.5%	29.1%
- Biomass	ktoe	174	254	275	279	n.a.	8.2%	n.a.
- Wastes	ktoe	202	270	289	404	n.a.	12.3%	n.a.
Renewable electricity generation (excl. hydro)	**GWh**	**967**	**1487**	**1852**	**2519**	**3675**	**17.3%**	**9.9%**
Percentage of total generation	%	1.3	1.9	2.3	3.0	4.1		
Geothermal	GWh	0	0	0	0	0	n.a.	n.a.
Solar, Wind, Wave, Tide	GWh	51	148	319	439	800	43.2%	16.2%
Biomass and Wastes[2]	GWh	916	1339	1533	2080	2875	14.6%	8.4%
- Biomass	GWh	0	115	237	270	825	n.a.	32.2%
- Wastes	GWh	916	1224	1296	1810	2050	12.0%	3.2%
Renewable TFC (excl. hydro)	**ktoe**	**176**	**233**	**223**	**221**	**400**	**3.8%**	**16.1%**
Percentage of TFC	%	0.3	0.4	0.4	0.4	0.6		
Geothermal	ktoe	0	0	0	0	0	n.a.	n.a.
Solar, Wind, Wave, Tide	ktoe	2	3	4	4	0	17.3%	n.a.
Biomass and Wastes[2]	ktoe	174	230	220	216	400	3.7%	16.6%
- Biomass	ktoe	174	230	220	216	400	3.7%	16.6%
- Wastes	ktoe	0	0	0	0	0	n.a.	n.a.
Hydro TPES	ktoe	10	10	8	7	26	-6.5%	39.4%
Hydro electricity generation	GWh	120	120	88	80	300	-6.5%	39.2%
Percent of total generation	%	0.2	0.2	0.1	0.1	0.3		

Notes:

1. Annual Growth Rate

2. Including Animal Products and Gases from Biomass

NEW ZEALAND

OVERVIEW OF RENEWABLE ENERGY POLICY

Hydro and geothermal resources already contribute significantly to New Zealand's energy supply and produce the majority of its electricity. Total non-hydro renewables (mainly geothermal and biomass) account for 14.8% of New Zealand's energy supply – compared to an IEA average of 3.9%.

The Government released its Renewable Energy Policy Statement in June 1993. The key objective of this policy is to ensure "the continuing availability of energy services at the lowest cost to the economy as a whole consistent with sustainable development." There are no specific quantitative targets or plans for future renewable energy use in New Zealand. However, government forecasts indicate that renewable energy supply will increase by around a fifth in 2010 compared with 1996, mainly due to increased electricity generation from wind, combustible renewables and wastes, and geothermal.

New Zealand is one of the few IEA countries which does not provide either output or capital subsidies to promote renewable energies, although some fiscal incentives are in place (in the form of import duty exemptions). Other government support includes identifying and responding to non-financial barriers to development, and identifying R&D priorities related to new and emerging renewable technologies. In addition, market reforms within the electricity sector have also included measures aiming to promote renewable electricity.

POLICIES

The Government produced a policy statement in 1993 stating its support for renewable energy and its renewables objective, which is to facilitate development of cost-effective renewable energy consistent with the Government's broader energy policy framework. That framework is essentially a commitment to ensuring competitive lowest-cost energy services while maintaining emphasis on the mitigation of environmental impacts from energy extraction, production and use. Other important policy drivers include: increasing energy security and diversity; maintaining access to energy services, particularly in remote areas; minimising the environmental effects of energy use, and providing export opportunities.

There are no quantitative targets for future renewable energy use in New Zealand, although their role in meeting energy needs is viewed as becoming increasingly important. This partly reflects the Government's view that these technologies are today at various stages of economic and/or technical development and that likely timing and progress of future development favours a flexible policy approach. Promotion of renewable energy is cited in New Zealand's National Communication under the FCCC, although no quantitative assessment of the level of associated GHG mitigation is given for individual renewable energy measures.

The Energy Efficiency and Conservation Authority (EECA) is responsible for renewable energy policy implementation, largely through dissemination of information, while the Ministry of Commerce undertakes relevant policy analysis and provides advice on renewables. Renewable energy promotional measures are contained in an integrated 10-point energy efficiency strategy. This strategy calls for technology demonstration and increased information provision to help the commercialisation of hydro, geothermal, and wind applications, and for the formation of a renewable energy R&D strategy that will accelerate the application of new and emerging technologies in New Zealand.

Government initiatives that encourage the development of cost-effective renewable energy applications include:

- *The 1992 Electricity Act and Energy Companies Act*. This allows independent power producers to supply direct to a specific local market or customer, and requires energy companies to disclose financial information to assist potential suppliers with line and energy cost information. Electricity market reform has separated the bodies responsible for transmission and generation, and has increased competition within generation. These developments play an important part in promoting the development of renewable energy sources for electricity generation including new renewables, and interest in these sources of energy has been growing. New renewables are exempt from a cap on the construction of additional generating capacity by the dominant generator, ECNZ;

- The *Resource Management Act 1992*. This act aims to promote sustainable management of natural and physical resources. Renewable energies are encouraged via a requirement on local authorities to "promote the sustainable management of natural and physical resources".

- The national CO_2 GHG emission mitigation programme[53]. This encourages renewables via: voluntary agreements with large CO_2 emitters to increase renewables' use; increased funding for EECA, and removing barriers (such as import duty, lack of technical standards) to the implementation of wind energy.

- Resource surveys and feasibility studies.

Other measures in place for encouraging the use of renewable energy sources include the provision of specialised information to assist planners and developers with consent applications under the Resource Management Act. In addition, imports of larger wind turbines are exempt from import taxes.

---

53 This is outlined in *Climate Change The New Zealand Response*, Ministry for the Environment, September 1994

The Government is working on further initiatives in the following areas to facilitate the cost-effective development of renewable energy:

■ Identification of barriers to deployment and appropriate response actions. Both EECA and the Ministry of Commerce are charged with maintaining surveillance on the renewables sector to identify and address barriers;

■ Identification of further research priorities for new and emerging renewables;

■ Progressive development of the national CO_2 emission reduction action programme;

■ Examination of how environmental criteria can be better incorporated into the decision making of energy producers and users. For example, under the *Resource Management Act* (New Zealand's principal environmental legislation) consents for activities are the responsibility of local authorities. Local authorities, when examining applications for consents, are required to focus primarily on the adverse environmental effects of activities, and to ensure that these are either avoided, remedied or mitigated. Under the Government's current CO_2 reductions programme, if net CO_2 emissions are not returned by 2000 to their 1990 levels, then a carbon tax will be imposed.

Table 1 shows that hydro efficiencies, gas-fired generation and geothermal are the most cost-competitive new electricity generation sources in New Zealand.

Table 1
Cost of additional generation capacity (1997)

Generation type	Total unit cost (NZ cents[54] /kWh)	Potential availability (MW)	Estimated Supply (GWh/yr)
Hydro efficiencies in current system	4	100	700
Gas combined cycle	4.3	400	3100
Low cost geothermal	5.5	200	1650
Mid cost geothermal	6.6	200	1650
Low cost wind	6.8	200	850
Coal	7.6	no limit	no limit
Mid-cost wind	8.3	200	850
High cost geothermal	8.6	400	3300
High cost hydro	8.7	500	4200
High cost wind	9.6	400	1700
Distillate	15.7	no limit	no limit

Source: *New Zealand Energy Outlook: February* 1997, Ministry of Commerce

54 On average in 1997, I US$ = 1.513 NZ$

Electricity from geothermal, and wind sources are also projected to increase to 2000 and beyond.

Non-cost barriers to further deployment of renewables include uncertainties arising from the implementation of electricity reforms (unbundling of wholesale, transmission and supply markets and functions), as well as barriers resulting from lack of information. In addition, the sometimes entrenched attitudes on the part of some architects, and industry and energy can also be a factor. Information availability should increase as the government is working with the renewable energy industry in order to produce documentation and seminars on renewable energy.

Renewable R&D forms a large share in total energy R&D in New Zealand, standing at 1.54 M USD (42% of energy R&D expenditure) in 1996. Wind and solar applications benefit from a higher level of funding, while funding for geothermal energy has dropped. Biomass R&D increased significantly in 1996 and focused on biogas from waste and solid biomass (forest waste and firewood). This shift in R&D funding reflects the government's renewable energy priorities. Moreover, the total amount of R&D funding is expected to increase: although just over 2% of the Public Good Science Fund (PGSF) was allocated to energy research in 1995/96 this is expected to increase substantially.

STATUS OF RESOURCE EXPLOITATION

Renewables are an important energy source in New Zealand. Hydro and geothermal electricity combined generated 77% of New Zealand's electricity in 1996. Government estimates depict this rising further as new renewables, and higher cost geothermal and hydro, become more competitive. Non-hydro renewable energy in New Zealand's energy supply was 2.4 Mtoe in 1996 (15% of TPES), one of the largest non-hydro renewables penetration in the IEA, mainly used for electricity generation. Around a third of renewable TPES is biomass, used mainly for domestic and industrial heating. Geothermal energy also provided 0.3 Mtoe of heat directly in TFC (used in household and industrial heating).

Government forecasts project renewable energy use increasing by around a fifth in 2010 compared with 1996, mainly due to increased electricity generation from geothermal sources (around 287 MW additional capacity over 1996 – or around 75% of low cost geothermal potential) followed by hydro (additional 358 MW or around 80% of the low cost hydro efficiency potential) and wind (additional 70 MW, or 25% of low cost wind potential). Total additional renewable generating capacity in 2010 is projected to be 960 MW.

Biomass

Supply of biomass has oscillated around 0.6 Mtoe since the beginning of the decade. This is made up predominantly of vegetal waste (used in industry) and wood (used in the residential sector). Electricity generation from biomass was 539 GWh in 1996, from three plants in the North Island.

Waste

In 1996, 55 ktoe of energy was recovered from industrial waste. Of this, 49 ktoe was used to generate electricity and 6 ktoe was used by industry for heating.

Wind

New Zealand's wind potential was estimated in 1993 at 12,656 GWh/yr (around half of current total electricity generation). The ECNZ wind turbine installed in Wellington has provided 1 GWh in 12 months from the 225 kW turbine. A number of interested parties, including a number of electricity distribution companies, are currently examining the potential for wind energy at various sites around New Zealand. New Zealand's first commercial wind farm, which has a capacity of 3.5 MW, is located at Haunvi and was commissioned in June 1996. Central Power, a Palmerston North based electricity distribution company, has committed to the development of a 32 MW wind farm to be commissioned in 1998. Government estimates indicate that installed wind capacity will be at least 70 MW in 2010.

Solar

Data not available, although both solar heating and PV applications have achieved commercial operation in the country.

Hydro

Hydro capacity and output has risen steadily from 18.9 TWh generated from 3952 MW in 1980 to 25.7 TWh from 5120 MW installed capacity in 1996. The majority of hydro capacity is from plants over 100 MW: small systems < 10 MW account for 118 MW. Hydro currently provides New Zealand with around 71% of electricity needs. No autoproducer hydro capacity is reported, although a number of small stations operated by power companies are connected to the grid.

Geothermal

In 1996, geothermal electricity generation provided 2091 GWh (around 5.8% of electricity generation) from 260 MW installed capacity. In addition, some geothermal energy is used directly (0.3 Mtoe in 1996). Generating capacity increased by 77 MW in 1997, with the commissioning of two units at Taupo. Capacity growth is expected to continue after the turn of the century, with government projections of 545 MW by 2010, when geothermal's contribution to electricity generation is estimated to rise to over 10%.

Heat pumps, Ocean, Wave, Tidal

Heat pump data is not available. No ocean, wave and tidal data is reported.

Table 2
Trends in renewable energy supply and use

	Unit	1990	1992	1995	1996	2000	1990-1996 (%)[1]	1996-2000 (%)[1]
Renewable TPES (excl. hydro)	**ktoe**	**2828**	**2945**	**2369**	**2405**	**3472**	-2.7%	9.6%
Percentage of TPES	%	20.0	19.6	15.3	14.8	20.1		
Geothermal	ktoe	2173	2269	1679	1698	2492	-4.0%	10.1%
Solar, Wind, Wave, Tide	ktoe	0	0	0	1	1	n.a.	13.5%
Biomass and Wastes[2]	ktoe	655	676	690	707	979	1.3%	8.5%
- Biomass	ktoe	601	622	635	651	n.a.	1.3%	n.a.
- Wastes	ktoe	54	54	55	55	n.a.	0.4%	n.a.
Renewable electricity generation (excl. hydro)	**GWh**	**2860**	**2962**	**2779**	**2881**	**3227**	0.1%	2.9%
Percentage of total generation	%	8.9	9.3	7.8	8.0	8.9		
Geothermal	GWh	2210	2272	2049	2091	2314	-0.9%	2.6%
Solar, Wind, Wave, Tide	GWh	0	0	1	7	7	n.a.	0.0%
Biomass and Wastes[2]	GWh	650	690	729	783	906	3.2%	3.7%
- Biomass	GWh	410	450	489	539	n.a.	4.7%	n.a.
- Wastes	GWh	240	240	240	244	n.a.	0.3%	n.a.
Renewable TFC (excl. hydro)	**ktoe**	**782**	**830**	**845**	**838**	**1067**	1.2%	6.2%
Percentage of TFC	%	8.2	8.1	8.0	7.9	8.8		
Geothermal	ktoe	272	315	323	315	348	2.4%	2.6%
Solar, Wind, Wave, Tide	ktoe	0	0	0	0	0	n.a.	n.a.
Biomass and Wastes[2]	ktoe	510	515	522	524	719	0.4%	8.3%
- Biomass	ktoe	504	509	516	517	709	0.4%	8.2%
- Wastes	ktoe	6	6	6	6	10	1.7%	12.2%
Hydro TPES	ktoe	2007	1774	2344	2215	2137	1.7%	-0.9%
Hydro electricity generation	GWh	23340	20631	27259	25750	24854	1.7%	-0.9%
Percent of total generation	%	72.3	64.5	76.8	71.2	68.6		

Notes:

1. Annual Growth Rate

2. Including Animal Products and Gases from Biomass

NORWAY

OVERVIEW OF RENEWABLE ENERGY POLICY

Norway has abundant resources of gas, oil and hydropower, and exports around eight times the amount of energy it consumes. Norway is one of the most electricity-intensive economies of the IEA, and electricity (over 99% of which is hydropower) accounted for 44% of TPES in 1995 (this dropped to 38% in 1996 due to low rainfall that year). Biomass contributes a further 5% to TPES, mainly in the domestic sector or in industry. Until recently, the pursuit of new renewables was extremely limited.

However, the environmental consequences of fossil fuel consumption, the need for greater energy flexibility and greenhouse gas commitments agreed to at Kyoto are seen by the government as important reasons to develop new alternative sources of energy (mainly wind electricity and biomass heat, as well as heat pumps). Given the overwhelming predominance of hydropower in electricity generation, the scope for CO_2 reductions in the electricity sector is lower than in other IEA countries. This results in a high marginal cost for emission reductions, and energy-related CO_2 emissions have in fact risen since 1990 and are projected to continue doing so. However, Norway was one of the first IEA countries to introduce a carbon tax, initiated in 1991 as a first step towards a comprehensive national climate policy. While this encourages renewables indirectly by discouraging fossil fuels (although air and sea transport fuels and process coal use in industry are exempt from the tax), new renewables are also encouraged directly through economic incentives, R&D, education/information campaigns and voluntary agreements.

The new government recognises that in order to increase use of new renewable energy sources there is a need for additional political measures and more favourable conditions for renewables than in the past. In April 1998 the government proposed new incentives for new renewables to the Storting (parliament), such as as exemptions from investment taxes for wind power, biomass systems and heat pumps. A support scheme for the production of wind power corresponding to 4-5 øre/kWh was also proposed.

POLICIES

Economic and environmental concerns are both important in Norwegian energy policy: although domestic energy use is less dependent on fossil fuels than many other IEA countries because of Norway's intensive use of hydro electricity, oil and gas exports generate over 15% of the country's income. Norway is one of three IEA countries to have Kyoto commitments allowing for a slight increase in greenhouse gas emissions, but containing emissions from the petroleum sector would be extremely challenging, given its economic importance and the currently low levels of per capita emissions from the energy sector in general, and projections in Norway's second National Communication[55] indicate that CO_2 emissions from this sector are expected to increase substantially by 2010.

The previous Government's *Long-Term Programme for 1998-2001* stated that the domestic use of electricity in a "normal year" should be based on renewable energy sources. Although hydropower generates almost all power in Norway at present, room for expansion is limited. This aim therefore provides an impetus for the development of electricity from other renewables, particularly wind and biomass. The development of non-hydro renewables would also help to diversify Norway's electricity supply sources, and therefore reduce supply vulnerability in dry years: at present, requirements for electricity imports are met by Sweden and Denmark. (However, the second National Communication projects CO_2 emissions from electricity generation to increase to 2 Mt/year by 2000, implying a diversification into gas-fired generation. This may not be contradictory to the stated aim of basing domestic electricity supply on renewables if gas-fired electricity is used solely for export. Nevertheless, it is unclear whether or not the new government will allow plans for gas-fired stations to proceed.)

The Ministry of Petroleum and Energy is in charge of policies to promote renewable energy use, and the Water Resources and Energy Administration (NVE) has the responsibility for accomplishing these policies. The combined budget for renewable energy promotion and energy efficiency has increased rapidly from 91 M NKR in 1997 to 157 M NKR in 1998. This is mainly disbursed via grants to producers of up to 50% in the R&D or pilot phase. Other means used for promoting new renewables are information campaigns (see below). However, the capital subsidies that were in place in the early 1990s for household PV systems have been stopped. Recent government proposals to encourage increased use of wind power, biomass, and heat pumps include investment tax rebates.

NVE uses information, training, education and technological dissemination to increase the awareness and practical skills regarding energy use, to facilitate approaches by individual consumer to consultants, contractors and producers of energy-efficient technologies. In addition, support is given at the national and regional level to businesses which offer energy-efficient technologies (services, solutions and products) so that these reach the level of market introduction as soon as possible. Regional energy efficiency offices have been set up in co-operation with electricity utilities, and are financed through a 0.2-0.3 øre[56] per kWh surcharge on consumer bills. The means used to promote renewables in the building and industry sectors are based on voluntary agreements and the use of existing networks in the different sectors.

---

55 *Norway's second national communication under the Framework Convention on Climate Change,* Ministry of Environment - Norway, Oslo, April 1997

56 On average in 1997, 7.07 NKR = 1 US $.

Since electricity supply is currently over 99% hydro-based, the scope for increased use of renewable energy in Norway is dependent on renewable applications that can substitute for oil and gas consumption, and continuing use of hydropower (or other renewables) for electricity generation. However, only a minor part, approximately 5.1 TWh, of the total remaining hydro potential is likely to be constructed in the near future: despite the economic importance of electricity exports, official Norwegian policy is that no pressure should be put on exploitation of water resources.

In the industrial sector, especially the pulp and paper industry, biomass and electricity have replaced some use of mineral oils as a result of measures introduced in the 1980s to regulate industrial emissions. High waste disposal costs also encourage this trend. The introduction of the carbon tax has also helped to encourage biomass and electricity use. However, the consumption of oil and other fuels varies depending on the spot price for electricity and other factors.

There are no general measures to promote biofuels for transport in Norway. One project in the region of Hadeland, which aims to produce biofuel from rape seed, has been subsidised. The Borregaard industries (pulp and paper) are exporting their ethanol to Sweden for use in buses in Stockholm.

Norway's renewable energy R&D budget has fallen sharply over the 1990s from 8 M USD in 1993 to 4.7 M USD in 1996 (10.4% of overall energy R&D expenditure). The majority of activity is focused on large scale hydropower (2.8 M USD), although other renewables (biomass, wind, solar and wave) have been funded since 1978. In 1996, approximately 0.8 M USD was spent on biomass, 0.6 M USD on solar applications and 0.4 M USD on wind under the NYTEK programme. The aims of R&D are the development of new renewable resources as realistic alternatives in long-term energy supply as well as exploring possibilities for commercial development of the relevant technology.

The government introduced guidelines for methane recovery from landfills in order to increase methane extraction. This has led to small amounts of methane capture and use (reflected in the municipal solid waste statistics).

The cost of renewable energy is judged as the most significant barrier to greater exploitation. The Norwegian Government estimates that the average cost for new hydropower is 22 øre/kWh, although the actual cost varies significantly from site to site. Heat produced from biomass costs a similar amount, whereas wind-based electricity costs 25-45 øre per kWh, and solar energy is even more costly at 40 øre or more per kWh (heat). However, siting problems are also judged an important barrier to increased use of wind, and lack of information on solar energy and biomass hinders greater use of these energy sources.

The marginal cost of carbon emission reductions in Norway is considered high, for reasons outlined above. For this reason the further investigation of policy concepts as joint implementation and emissions trading is of great importance for Norway.

STATUS OF RESOURCE EXPLOITATION

The importance of non-hydro renewable energy in Norway was 5.2% of TPES in 1996. This is predominantly biomass used directly in the industrial and residential sectors, although there is a small amount of biomass use for electricity and heat, wind-generated electricity and some rooftop PV installations. On top of this, hydropower generally contributes over 38% of Norway's TPES, and almost all of its electricity.

Biomass

Norway has large biomass reserves, with plenty of room for expansion despite a significant use of biomass at present. Total final consumption of fuelwood, black liquor and waste was 1.02 Mtoe in 1996, divided between industry (0.40 Mtoe) and private households (0.63 Mtoe). If biomass use were to increase, it would probably be from increased biomass use is industry, particularly those companies with access to biomass such as in the forestry industry. Such industries have the added incentive of simultaneously reducing their waste flows, which would increase profitability. Biomass is a less economically attractive option in other sectors due to its cost. However, the latest official forecasts do not suggest that a significant increase is likely.

Waste

Industrial waste contributed 3.6 ktoe to TPES in 1996. Municipal waste contributed a further 117 ktoe in 1996. The energy recovered from industrial waste is sold as heat whereas that from municipal waste is used to generate small amounts of electricity as well as heat (57 GWh of electricity and 3993 TJ of heat in 1996).

Wind

Wind energy is seen in Norway as the most developed 'new' renewable (i.e. excluding biomass and hydropower), and is reported as contributing 0.6 ktoe to Norway's TPES in 1996. Wind capacity has remained at under 4 MW (installed at sites on the west coast) since 1993, although the state-owned generator, Statkraft, has estimated that Norway has significant windpower potential of approximately 14 TWh/year. However, capacity should increase over the next few years, as there are plans for 25 MW to be installed by 1999. Although the cost of wind generation is estimated at NKr 0.25-0.45/kWh (compared to hydro's NKr 0.22/kWh) windpower could become competitive in Norway either due to cost decreases in the large turbines being installed, or in dry years, when the spot price of electricity rises rapidly. The new government has signalled that 2-4 TWh of wind power should be the objective within 10 years. (For example, 1996 was a very dry year and the electricity spot price rose above NKr 0.3 for approximately 5 weeks.) However, wind generation is likely to remain small-scale for the foreseeable future.

Hydro

In 1995, hydropower generated 121 TWh from approximately 26 GW of installed hydro capacity. This was over 99% of Norway's electricity production. Production was lower in 1996 (104 TWh) as this was a dry year. Small hydro stations (<10 MW) generate approximately 3.8 TWh annually from a total capacity of 880 MW. Around 12% of installed capacity is reported as autoproduction. However, growth in hydro production will be limited in future, due to environmental constraints on the building of new hydro plants: capacity is expected to grow less than 300 MW between 1995 and 2000.

Solar

Generation of 3 GWh from 4 MW installed capacity was reported for 1993. One PV-diesel installation has been built, mainly with public funds, to support the electricity to one family home. This is a demonstration project to get experience with a PV-diesel hybrid, which offers a possible solution for cost-effective energy supply in some isolated areas of Norway.

Wave

Norway focussed much of its earlier efforts in renewables R&D during the late 1970s and early 1980s on wave power, due to Norway's long coastline and good wave conditions. This resulted in two demonstration projects near Bergen in the mid-1980s. However, the current status of wave technology and the price of alternative energy sources means that wave power will not be a realistic power source for the foreseeable future.

Heat pumps

In Norway there is a particular potential for residential and commercial space heating. Both the population and industry of Norway are spread over the whole country, so the costs of installing a district heating network would be high due to transport of heat over long distances. The use of heat pumps based on electricity can represent an appropriate way to meet Norwegian heat demand in an efficient manner.

A heat pump programme ran until 1994, when 17,000 units had been installed. Heat extracted from the surroundings was estimated at 2 TWh. NVE's continuation of the heat pump programme rests on increased information dissemination, demonstrations and R&D.

Table 1
Trends in renewable energy supply and use

	Unit	1990	1992	1995	1996	2000	1990-1996 (%)[1]	1996-2000 (%)[1]
Renewable TPES (excl. hydro)	**ktoe**	**1002**	**1015**	**1178**	**1212**	**1302**	**3.2%**	**1.8%**
Percentage of TPES	%	4.7	4.5	5.0	5.2	5.5		
Geothermal	ktoe	0	0	0	0	0	n.a.	n.a.
Solar, Wind, Wave, Tide	ktoe	0	0	1	1	2	n.a.	35%
Biomass and Wastes[2]	ktoe	1002	1015	1177	1211	1300	3.2%	1.8%
- Biomass	ktoe	903	906	1059	1091	n.a.	3.2%	n.a.
- Wastes	ktoe	99	108	118	121	n.a.	3.3%	n.a.
Renewable electricity generation (excl. hydro)	**GWh**	**242**	**251**	**323**	**338**	**285**	**5.7%**	**-4.2%**
Percentage of total generation	%	0.2	0.2	0.3	0.3	0.2		
Geothermal	GWh	0	0	0	0	0	n.a.	n.a.
Solar, Wind, Wave, Tide	GWh	0	3	10	7	25	n.a.	37.7%
Biomass and Wastes[2]	GWh	242	248	313	331	260	5.4%	-5.9%
- Biomass	GWh	184	206	265	274	n.a.	6.9%	n.a.
- Wastes	GWh	58	42	48	57	n.a.	-0.3%	n.a.
Renewable TFC (excl. hydro)	**ktoe**	**878**	**903**	**1022**	**1052**	**1200**	**3.1%**	**3.3%**
Percentage of TFC	%	4.9	5.1	5.3	5.4	6.1		
Geothermal	ktoe	0	0	0	0	0	n.a.	n.a.
Solar, Wind, Wave, Tide	ktoe	0	0	0	0	0	n.a.	n.a.
Biomass and Wastes[2]	ktoe	878	903	1022	1052	1200	3.1%	3.3%
- Biomass	ktoe	878	903	1022	1052	n.a.	3.1%	n.a.
- Wastes	ktoe	0	0	0	0	n.a.	n.a.	n.a.
Hydro TPES	ktoe	10418	10034	10435	8911	9830	-2.6%	2.5%
Hydro electricity generation	GWh	121145	116672	121343	103619	114220	-2.6%	2.5%
Percent of total generation	%	99.6	99.6	99.4	99.2	99.6		

Notes:

1. Annual Growth Rate

2. Including Animal Products and Gases from Biomass

PORTUGAL

OVERVIEW OF RENEWABLE ENERGY POLICY

Portugal's rapid energy growth and heavy reliance on imported fuels has shaped its priorities for future energy development. These priorities are laid out in the 1994 *Energy Programme* created under the Community Support Framework, and improved energy security and diversity are high on the agenda. The main lines of the *Energy Programme* are enhanced electricity and gas services, energy efficiency and renewable energy (particularly renewable electricity) promotion. The government has a short-term target for small hydro, wind and biomass generating capacity that is likely to be met on time, and meeting this target may help in slowing the declining importance of renewable energy in Portugal's energy mix in the future. To date, the renewables portion of the *Energy Programme* has had most success with small (<10MW) hydro plants and wind farms.

The increased capacity of renewable electricity systems has been brought about by a mix of measures, notably generous capital investment subsidies or no-interest loans and a guaranteed market and price for the renewable electricity produced. It is probably this last measure that has succeeded in increasing investment in renewable electricity systems, as the cost to consumers of electricity has fallen sharply over the last decade, e.g. real industrial electricity prices fell 30% between 1987-1996. Because of this fall, payment terms for new renewable electricity were improved in November 1995.

Non-hydro renewable energy contributed 6.1% to Portuguese TPES and 3.0% to electricity generation in 1996, compared to IEA averages of 3.9% and 2% respectively. The absolute amount of renewable energy used is likely to grow slightly in the short-term, with the commissioning of wind and biomass electricity generating systems, although the importance of renewable energy is unlikely to change significantly.

POLICIES

Portugal's *Energy Programme* was defined most recently in 1994 in the Decree number 195/94. This programme was designed to achieve a number of objectives, including a reduced dependence on energy (and especially oil) imports; a reliable energy supply at a reasonable cost; increased energy conservation and reduced environmental impacts of energy use. Part of the *Energy Programme* is devoted to encouraging the increasing use of renewable energies (mainly renewable electricity) not only to increase energy self-sufficiency from its currently low level, but because of the positive effects that increased renewable energy use has on the environment and on regional development. The short-term target for renewable energy is to have 90 MW small hydro, 70 MW wind and 10 MW biomass capacity in place by 2000.

Portugal uses a range of measures to promote renewable energy including:

- measures to encourage renewable electricity production (guaranteed markets and favourable prices for renewable electricity and CHP);

- direct capital investment subsidies;

- other market stimulation incentives, e.g. no-interest loans;

- information/education campaigns; and

- R&D.

The renewable part of the *Energy Programme* is run by the General Directorate of Energy under the Ministry of Economics, with almost a third of the total estimated financial requirements of 182 M contos[57] provided by the EU programme FEDER. The form of support for renewable energy projects under the *Energy Programme* depends on the nature of individual projects, with projects benefitting from a grant of up to 60% of eligible costs if they are demonstration projects, up to 50% if they are dissemination (commercialisation) projects, and loans (that could possibly be transformed into grants if the project is considered "excellent") up to 40% for projects aiming to increase the deployment of mature technology. The subsidies per project are capped at 50 000 contos (50 M ESC) except for CHP systems, where the cap is 150 000 contos. The exact level of support for an individual project varies depending on its size (projects under 10 MW receive most help), and its regional and environmental impacts.

The Portuguese electricity supply industry is undergoing a period of change, with the introduction of natural gas as well as, to a lesser extent, the increased use of renewable energies. In addition to the capital subsidies described above, the Government encourages independent renewable power generation allowing any organisation (but not individual households) to qualify as an independent producer with the right to supply the grid up to 10 MW at regulated prices (set out in Law 313 in 1995). The prices vary depending on when the electricity is produced, with a premium for peak or shoulder rate electricity. Average prices paid were estimated at 10.8 ESC/kWh in 1996. For systems larger than 10 MW, incremental electricity production (i.e. from all but the first 10MW capacity) will be paid for at "avoided cost" rates for 15 years. The purchase power obligation and technical conditions for installing connections to the national grid are set out under Decree Law 189 (1988), and include provisions for a guaranteed market for renewable electricity at favourable prices. This produced significant investment in small hydro in the early 1990s, and in wind in the mid 1990s.

---

57 In 1997, 1 $ US = 175.2 ESC. 1 contos = 1000 ESC

Purchases of renewable energy equipment (such as solar panels for residential use) benefit from the reduced VAT rate of 5% under the 1992 Budget Law. It is also possible to deduct the investment cost in renewable end-use technology from personal taxable income (subject to a ceiling).

Public energy R&D is managed through the Institute for Industrial Engineering and Technology (INETI). Significant reorientation of support away from energy conservation and fossil fuels towards renewable R&D took place in 1993-94 with renewable sources receiving a rapidly growing portion of the overall energy R&D budget. Overall spending on renewable energy was increased by 44% in 1994 (to 2.1 M USD) while total energy R&D spending remained roughly the same. The main thrust of R&D was solar (mainly PV) accounting for over half the total renewable R&D. Biomass received the next largest share of funds (22%) with the remainder spent on small hydro, geothermal, wind and ocean energy.

According to the Energy Directorate of the Ministry of Economy (DGE), increased use of some passive solar design features have been achieved through regulations on the characteristics of the thermal behaviour of buildings.

The government is also working to increase information dissemination and education about renewable energies in Portugal, with publicly available information on existing subsidies and how to apply for them. In addition, a biomass centre for energy has operated since 1989, coordinating and promoting demonstration projects in the areas of biomass production, collection transformation and use.

STATUS OF RESOURCE EXPLOITATION

Non-hydro renewable energy provided 6.1% of Portugal's TPES in 1996, down from 7.1% in 1990. Biomass and wastes account for 95% of all non-hydro renewable use. There are different trends within different renewable energies, with increases in renewable electricity generation, and a slight decrease in the direct use of renewables (biomass) mainly in the industrial or residential sectors. In 1996, large and small hydro contributed a further 6.6% to Portugal's TPES and 42.9% to total electricity generation, producing 14761 GWh.

Biomass

Biomass is by far the largest source of renewable energy in Portugal, at 1.1 Mtoe in 1996, accounting for 5.7% of TPES in the same year. The vast majority of this, 0.96 Mtoe, was used directly either in the residential sector (wood) or in the industrial sector (wood wastes). The remainder was used to generate 960 GWh of electricity, produced solely from CHP plants.

Biomass electricity production has increased rapidly over the last few years, up 39% since 1990, largely due to the incentives available for CHP partially fired by biomass. Biomass electricity generation is likely to increase in the short term, given the continuation of the present incentives, and with a 10 MW electricity plant using forest residues under construction.

Waste

No energy recovery from municipal waste is reported to the IEA.

Wind

Wind capacity has risen from 1 MW in 1990 to 18 MW in 1996, and will continue to rise sharply as a further 42.4 MW are either under construction or planned to be operational before 1999. The favourable regulations covering installations up to 10 MW means that none of the planned windfarms are larger than 10 MW. This regulation could have ultimately slowed down the takeup of windpower if it had not been removed by Decree 313/95. Potential for total capacity has been estimated at 400 MW, although this would be reduced to 240 MW and approximately 2% of total electricity generation if siting and other environmental restrictions on wind farms are taken into account. The planned capacity additions will ensure that the 21 GWh produced from wind in 1996 will grow substantially in the short term.

Hydro

In 1996, hydro contributed 42.9% (14761 GWh) to electricity generation. Installed capacity rose from 2516 MW in 1980 to 4428 MW in 1996. Small hydro capacity was estimated at 240 MW in 1996 – a significant expansion over the 1990s. The additional potential for small hydro capacity is estimated at approximately 100 MW to 2000 and an additional 250 MW above that in the longer term. Total hydro capacity is forecast to increase to 5489 MW by 2010. Deliveries to the grid from independent hydropower producers increased from 336 GWh in 1995 to 520 GWh in 1996.

Solar

Portugal's Energy Department (DGE) estimates that 200,000 m² of solar heaters have been installed by the end of 1996, as well as some pilot PV plants (400 kW). However, electricity generation from PV has not taken off as fast as that from other renewables partly because individuals are excluded from benefitting from the favourable buyback rates described in the previous section. This therefore precludes benefits for distributed PV generation, and significant expansion of PV is unlikely unless this regulation is changed.

Table 1
Trends in renewable energy supply and use

	Unit	1990	1992	1995	1996	2000	1990-1996 (%)[1]	1996-2000 (%)[1]
Renewable TPES (excl. hydro)	**ktoe**	**1163**	**1083**	**1155**	**1162**	**1340**	**0.0%**	**3.6%**
Percentage of TPES	%	7.1	6.1	6.0	6.1	6.4		
Geothermal	ktoe	3	4	37	43	30	52.4%	-8.7%
Solar, Wind, Wave, Tide	ktoe	11	14	17	17	50	7.9%	30.0%
Biomass and Wastes[2]	ktoe	1148	1065	1101	1102	1260	-0.7%	3.4%
- Biomass	ktoe	1148	1065	1101	1102	n.a.	-0.7%	n.a.
- Wastes	ktoe	0	0	0	0	n.a.	n.a.	n.a.
Renewable electricity generation (excl. Hydro)	**GWh**	**695**	**892**	**1047**	**1031**	**1430**	**6.8%**	**8.5%**
Percentage of total generation	%	2.5	3.0	3.2	3.0	3.6		
Geothermal	GWh	4	5	42	49	30	51.8%	-11.5%
Solar, Wind, Wave, Tide	GWh	2	5	17	22	180	49.1%	69.1%
Biomass and Wastes[2]	GWh	689	882	988	960	1220	5.7%	6.2%
- Biomass	GWh	689	882	988	960	n.a.	5.7%	n.a.
- Wastes	GWh	0	0	0	0	n.a.	n.a.	n.a.
Renewable TFC (excl. hydro)	**ktoe**	**1010**	**929**	**968**	**975**	**1010**	**-0.6%**	**0.9%**
Percentage of TFC	%	8.0	7.0	6.7	6.4	6.3		
Geothermal	ktoe	0	0	1	1	0	n.a.	n.a.
Solar, Wind, Wave, Tide	ktoe	11	13	15	16	30	6.1%	17.8%
Biomass and Wastes[2]	ktoe	999	915	951	959	980	-0.7%	0.5%
- Biomass	ktoe	999	915	951	959	n.a.	-0.7%	n.a.
- Wastes	ktoe	0	0	0	n.a.	n.a.	n.a.	n.a.
Hydro TPES	ktoe	788	400	717	1269	950	8.3%	-7.0%
Hydro electricity generation	GWh	9157	4646	8343	14761	11000	8.3%	-7.1%
Percent of total generation	%	32.3	15.7	25.2	42.9	27.4		

Notes:

1. Annual Growth Rate

2. Including Animal Products and Gases from Biomass

185

Geothermal

Portugal has two geothermal electricity plants in the Açores with a total capacity of 8.5 MW. Generation was 49 GWh in 1996, up from 4 GWh in 1990.

Ocean/Tidal/Wave/ Heat pumps

None reported to IEA. However, wave energy resource mapping surveys have been undertaken, and INETI report that a 500 kW wave plant is under construction in the Açores.

SPAIN

OVERVIEW OF RENEWABLE ENERGY POLICY

The government has ambitious plans for the development of renewables. The *Energy Savings and Efficiency Plan* (PAEE)[58] aims to increase use of renewables by 1.1 Mtoe by the year 2000 and to increase the contribution of non-hydro renewables in electricity generation from 0.5% in 1990 to 1.4% in 2000. Regional government promotion, EU funding and private sector investment promise to help achieve the government's objective. The two main thrusts of renewables promotion in Spain are favourable electricity tariffs to autoproducers and capital subsidies under the PAEE programme.

The government undertakes regular evaluations of progress under the PAEE, and significant progress towards its aims has been achieved, with over 43% of the PAEE objectives met by the end of 1996. This rises to almost 71% if plant under construction is included. The cost of the PAEE programme is high: the cost to 1996 was 129.5 billion Ptas[59] of private capital and 23.1 billion Ptas (182 M USD) of public funding – about half of the estimated total funding at the beginning of the plan. (This does not take into account expenses incurred under the favourable buy-back rates). However, the success of the plan to date has allowed the government to cut back on the systems and regions eligible for capital subsidies, as well as on the subsidy amounts per project, and still be on course for achieving the overall target laid out at the start of the plan.

Private and public funding are directed towards those electricity-producing sources closest to commercialisation, i.e. small hydro and wind. (Electricity-producing technologies also benefit from a favourable purchase price of electricity). Public funding is also directed towards solar PV where costs of production are decreasing and where the objectives are less ambitious than for the other renewables. As a consequence, the objectives set for wind and solar PV in 2000 have been surpassed, small hydro is well advanced, whereas the increased use of urban waste, geothermal and solar thermal is much slower than anticipated. It is possible that these sources may require additional incentives if their target penetration is to be achieved or, conversely, that other non-financial barriers preventing deployment are addressed. Alternatively, a higher than originally anticipated use of wind and solar energy (at the expense of other renewables) could enable the government to meet its total renewable target.

---

58 This plan excludes large hydro.

59 In 1997, 1 $ US = 146.4 Ptas

POLICIES

The majority of Spain's energy needs are met by imported fossil fuels (mainly oil), with nuclear power an important source of national electricity production but constrained from further growth by a moratorium. This relative lack of indigenous energy reserves has shaped Spain's energy policy, which has security of low cost energy supply from a diverse range of domestic resources as its key objective. The country is currently halfway through its latest *National Energy Plan* (PEN) which runs from 1991-2000.

A number of national plans and sub-plans for renewable energy have been drawn up during the last decade. Promotion of renewable energy in Spain has been supported by the Spanish government under the PEN, through two Renewable Energy plans (PER 86-88 and PER 89-90) and currently under the *Renewable Energy Programme,* which is one of four sub-plans under the PAEE, itself included in the *National Energy Plan.* The government's strategy to promote increased use of renewables rests on two main approaches: favourable buy-back rates for electricity produced from facilities under 100 MW; and financial support via capital subsidies and loans for third party financing under the PAEE plan.

Responsibility for the promotion of renewable energy lies with the Institute for Energy Diversification and Conservation (IDAE), a state 'mid-autonomous' organisation under the supervision of the Ministry for Industry and Energy. Renewable energy is also promoted at the regional and local levels, e.g. the Andalusian government launched a 3 year programme for solar heating systems with a budget of 1 200 million Ptas. IDAE has also launched renewable programmes with the regions of Canarias, Aragon, Valencia and Murcia.

The emphasis of the renewables programme of the PAEE is on new electricity generating technologies which reduce dependence on fossil fuels, although use of renewable-based heat is also included. Biofuels for transport are excluded from the plan. Table 1 shows the PAEE targets and status at the end of 1996 of the different renewable energies covered under the plan.

The overall objective of the PAEE is to increase renewable energy production to 1.1 Mtoe/year by 2000, increasing renewable electricity production to 4.2 TWh/y and thermal energy to 0.5 Mtoe, comprising 85% increase from biomass, 9% from small hydro, a planned target for solar collector area for 2000 of 400,000 m^2/y and installed wind capacity of 168 MW.

The government has envisaged that the total financing required to achieve the target (approximately 334 billion Ptas at 1991 prices) will be mainly private capital, but with a total public commitment of 70 billion Ptas. By the end of 1996, total private funding for systems put in place and under construction amounted to 129.5 billion Ptas, and total public funding was 23.1 billion Ptas. The majority of public funding has been via subsidies from the Ministry of Industry, the regions, EU

Table 1
Targets for Renewable Energy Use in 2000

	Objective 1991-2000 (ktoe)	Completed as of 31/12/96 (ktoe)	% of the objective	Total funding to end 1996 (M Ptas)*
Small hydro	213	126	59.3	64852
Biomass	427	165	38.7	13140
Urban waste	446	169	37.9	28949
Wind	34.6	43.2	125	32101
Solar PV	0.39	0.55	141	10040
Solar Thermal	33.7	3.6	10.7	3494
Geothermal	10	0.4	4 4	66.5
TOTAL	1165	508.8	43.7	14940

Source: Ministry of Industry and Energy, IDAE

* Includes financing for non specified projects

programmes, the central administration and IDAE. Funds are directed mainly towards small hydro (42% of total), wind (21%) and urban waste (19%), and the majority of public funding is higher for technologies that are more expensive, such as solar PV. Remaining public funds were disbursed via loans from IDAE for third party financing, with small amounts for the promotion of and management of the programme.

The government estimates that the renewables section of the PAEE programme, including projects currently under construction, place Spain almost 60% of the way towards its 2000 target for total electricity generated from renewables, see Table 2. Wind and PV have exceeded their 2000 targets ahead of schedule, whereas use of municipal wastes has been slower than hoped. Progress towards increased use of renewables for non-electricity generating purposes, i.e. solar thermal and geothermal is also slow, and these projects have accounted for under 3% of total funding to date (perhaps because total incentives for these projects are lower as they do not benefit from favourable buy-back rates).

Investment has been a critical factor in Spain's achievements in the use of renewable energy technologies to date. An investment of 1 billion ECU in the period 1986-1991 will rise to a cumulative investment of 3.5 billion ECU (1990 prices) at the expiry of the current energy plan. Approximately 70% of the additional 2.5 billion ECU is earmarked for small hydro and MSW projects. Government spending on renewable energy sources rose by a quarter in 1994 to 3,122 million Ptas.

Table 2
Renewable Electricity Targets

		PAEE Target	Completed (end 1996)	% target met
Small Hydro:	Capacity (MW)	779	408	52.4
	Generation (GWh/y)	2474	1469	59.4
Municipal Solid Waste:	Capacity (MW)	239	67	28.2
	Generation (GWh/y)	1298	502	38.7
Wind:	Capacity (MW)	168	205	122
	Generation (GWh/y)	403	502	125
Solar PV:	Capacity (MW)	2.5	3.8	152
	Generation (GWh/y)	4.5	6.4	142
Total:	Capacity (MW)	1188	684	57.5
	Generation (GWh/y)	4179	2479	59.3

Source: MINER, 1997

Until 1995, IDAE concentrated its efforts on small hydro, but has since 1996 been investing more heavily in wind energy because the purchase price of electricity makes these projects more attractive. The IDAE can finance up to 100% of the eligible cost (directly linked to the cost of energy-producing equipment, but also including "energy audits") of a project, and is developing financing through joint ventures. State and regional subsidies finance less economically attractive projects. Every year a Decree sets the share of eligible costs that can be subsidised (see Table 3). This amount varies by project type as well as varying over time: for example, capital subsidies for wind systems are now only available in certain regions, while those for biomass projects have been strengthened.

There is an obligation for utilities to buy excess electricity from autoproducers at a price set by the Administration. A December 1994 Decree on electricity produced by autogenerators from hydro, co-generation and renewables, overturned a previous decree that pegged buy-back rates to the cost of small hydro generation. The new decree only applies to plants with a capacity lower than 100 MW (lower than 10 MW for hydro) and sets rates for both capacity and output credits (buy-back rates). Capacity credits are highest for waste incineration plants, whereas output credits are highest for wind and solar plants: 11.48 Ptas/kWh over a five-year period. Output credits for waste-generated electricity vary, depending on the size of the plant and the relative importance of any co-fired fossil fuel, but are lower than those for wind and solar electricity, and decrease yearly. Buyback rates for such plants vary from 9.19-9.92 Ptas/kWh in the first year – still significantly higher than the estimated average production cost from autoproducers of 8 Ptas/kWh. The buyback rate levels are also dependent on continuity of supply to avoid periodic surges in power sold to the grid.

Spain spent 16.3 M USD, or 20.9% of its total government energy R&D expenditure, on renewables in 1995. The majority, 12.2 M USD, was spent on solar energy (mainly solar thermal). Biomass research accounted for 3.2 M USD and wind accounted for 1 M USD: no R&D on hydro or geothermal was funded nationally.

Table 3

Maximum subsidies for renewable energy systems under the PAEE

	Subsidy (maximum percentage of eligible cost)		
	1997-1999*	1996	1995
Wind	- up to 30%** (some constraints on machine size, windfarm capacity)	- up to 30% (all systems)	- up to 30% for single turbines or small wind farms
Solar thermal	15000-35000 Ptas per m² for an installation of at least 30m²	- up to 35000 Ptas per m² for an installation of at least 40m² that is going to function for at least three years	- up to 27000 Ptas per m² for an installation >40m² that will function for at least three years
Solar PV	- up to 600 Ptas/Wp for grid-connected generation (<5 kW)	- up to 600 Ptas/Wp for grid-connected generation (end of line) - up to 800 Ptas/Wp for other grid-connected systems	- grid connected systems: up to 800 Ptas/Wp
	- isolated systems: up to 1200 Ptas/Wp	- isolated systems: up to 1600 Ptas/Wp	- isolated systems: up to 1600 Ptas/Wp
Biomass and wastes	- up to 30% for fuel production** and electricity generation	- up to 30% for projects substituting fossil or electricity use with the use of wastes	- up to 20%
(Only available for plants firing up to 10% of "conventional" fuel).	- up to 25%** for use of municipal solid waste - up to 20%** for biogas	- up to 15% for electricity generating projects - up to 15% for other projects (e.g. biogas)	- up to 15%
Hydro	Up to 5%** for systems under 10 MW	Island systems: up to 20% Improvements to mainland water supply: up to 10%	Island systems: up to 20% Improvements to mainland water supply: up to 10%

* Only certain regions of Spain are eligible for grants from 1997.

** Small and medium businesses ("PYMES") can obtain subsidies up to 10 percentage points higher.

STATUS OF RESOURCE EXPLOITATION

In 1996, non-hydro renewable energy sources accounted for 3.6 Mtoe i.e. 3.5% of TPES, just under the IEA average of 3.9%. The use of hydropower (mainly large hydro) is of a similar order of magnitude, at 3.4 Mtoe and 3.4% of TPES. Non-hydro renewable energy use in 1996 was almost exclusively (over 98%) made up of combustible renewables and wastes, although wind and solar energy use are growing fast. Electricity generation from municipal waste has also increased rapidly. However, total non-hydro renewable electricity still accounts for a very small percentage of total electricity generation.

Biomass

The 1991-2000 objective for biomass use is for an increase of 427 ktoe. However, the uptake of biomass projects has been slower than anticipated, with less than 40% of the government's target being met by the end of 1996. The use of wood and vegetal wastes has decreased over the 1990s, while that of black liquor and biogas has increased. Biomass-based generation has increased from 484 GWh in 1990 to 936 GWh in 1996. Provisional investment data for 1997 indicates that, if anything, installation rates of biomass systems are declining.

Waste

The *National Energy Plan* includes a target of 446 ktoe of energy from municipal solid waste in 2000. If this target is to be met, installations will have to be built at a greater rate, because the capacity and generation targets (see Table 2) were at less than 40% by the end of 1996. Even if installations under construction are included, production is only just over 50% of the target more than halfway through the plan. Generation estimates from industrial and municipal waste combined were 562 GWh in 1996: substantially up from 1994, when generation was 293 GWh.

Wind

Wind has been one of the government's success stories, and installed capacity has grown from 7 MW in 1990 to over 377 MW in November 1997 – more than double the government's target of 168 MW by 2000. Growth of installed wind capacity has been very rapid in the last few years: total capacity was 221 MW in 1996 and 115 MW in 1995. Generation from wind turbines was 175 GWh in 1994 and 364 GWh in 1996. This growth in wind energy is due to a number of factors, including the generous capital and output subsidies in place and the high potential of wind energy in Spain (estimated by the government to be 2.8 GW).

There is significant potential for continued expansion as potential for wind power has been estimated at 2.8GW. However, expansion may be limited or delayed by the restrictions imposed by current regulations on grid-connection. Expansion may also be limited if legislation regarding favourable buy-back rates is changed. Nevertheless, according to some national estimates, installed capacity could be as high as 750 MW by 2000.

Solar

PV capacity was estimated at 6.9 MW (generating about 12.2 GWh) in 1996, of which approximately 4 MW had been installed under the PAEE. The current level of expansion accelerated in the early 1990s, but has levelled off since 1994.

Hydro

Hydropower generated just under 40 TWh in 1996: substantially higher than generation in 1995 due to capacity increases. The majority of generation was from large hydro, but small hydro plants (<10 MW), which are encouraged under the PAEE plan produced over 3.2 TWh in 1995, and an estimated 5.3 TWh in 1996. Installed capacity in 1996 was 11.8 GW, and there is significant room for expansion for both large and small hydro schemes, although exploitation of some of the larger potential sites may be constrained by conflicts with water extraction policy regulations. The importance of hydropower in Spain's electricity mix has grown over recent years, from 12.1% in 1992 to 23.0% in 1996.

Geothermal

The energy-saving potential of Spain's geothermal resources located in the Canary Island is estimated at around 600 ktoe/yr. However, extremely limited interest has been shown in development of geothermal resources: only two projects producing 0.4 ktoe of heat have been initiated under the renewables portion of the PAEE plan, and no further plants are under construction. This is due to the delays in the construction of the most important of these geothermal plants in Madrid. Unless increased interest in geothermal energy sources is shown, the government's target for 2000 is extremely unlikely to be met.

Table 4
Trends in renewable energy supply and use

	Unit	1990	1992	1995	1996	2000	1990-1996 (%)[1]	1996-2000 (%)[1]
Renewable TPES (excl. hydro)	**ktoe**	**3392**	**3464**	**3563**	**3584**	**3500**	**0.9%**	**-0.6%**
Percentage of TPES	%	3.7	3.6	3.4	3.5	3.2		
Geothermal	ktoe	0	0	7	7	0	n.a.	n.a.
Solar, Wind, Wave, Tide	ktoe	2	4	49	57	0	79.2%	n.a.
Biomass and Wastes[2]	ktoe	3390	3459	3508	3520	3500	0.6%	-0.1%
- Biomass	ktoe	3297	3361	3219	3222	n.a.	-0.4%	n.a.
- Wastes	ktoe	93	98	289	299	n.a.	21.4%	n.a.
Renewable electricity generation (excl. hydro)	**GWh**	**714**	**774**	**1608**	**1874**	**2932**	**17.6%**	**11.8%**
Percentage of total generation	%	0.5	0.5	1.0	1.1	1.4		
Geothermal	GWh	0	0	0	0	0	n.a.	n.a.
Solar, Wind, Wave, Tide	GWh	20	52	285	376	432	63.1%	3.5%
Biomass and Wastes[2]	GWh	694	722	1323	1498	2500	13.7%	13.7%
- Biomass	GWh	484	500	814	936	1050	11.6%	2.9%
- Wastes	GWh	210	222	509	562	1450	17.8%	26.7%
Renewable TFC (excl. hydro)	**ktoe**	**2778**	**2823**	**2667**	**2667**	**2700**	**-0.7%**	**0.3%**
Percentage of TFC	%	4.5	4.3	3.7	3.7	3.4		
Geothermal	ktoe	0	0	7	7	n.a.	n.a.	n.a.
Solar, Wind, Wave, Tide	ktoe	0	0	25	25	n.a.	n.a.	n.a.
Biomass and Wastes[2]	ktoe	2778	2823	2636	2636	2700	-0.9%	0.6%
- Biomass	ktoe	2754	2799	2541	2541	n.a.	-1.3%	n.a.
- Wastes	ktoe	24	24	95	95	n.a.	25.7%	n.a.
Hydro TPES	ktoe	2186	1627	1988	3423	3100	7.8%	-2.4%
Hydro electricity generation	GWh	25414	18920	23112	39802	36600	7.8%	-2.1%
Percent of total generation	%	16.8	12.1	14.0	23.0	17.9		

Notes:

1. Annual Growth Rate

2. Including Animal Products and Gases from Biomass

SWEDEN

OVERVIEW OF RENEWABLE ENERGY POLICY

The objectives of Swedish energy policy are to secure short- and long-term electricity supply, as well as the supply of other energy, in a competitive manner. As such, energy policy aims to minimise negative impacts on health, the environment, and the climate while providing conditions for efficient energy use and cost-effective energy supply. Energy policy's role in sound economic and social development in Sweden is also important. In this context, the parliament passed a bill on Sustainable Energy Supply in June 1997.

Non-hydro renewable energy use in Sweden has significantly expanded since 1990, and accounted for 14.5% of TPES in 1996 (more than triple the IEA average). In addition, hydropower contributed 8.4% to TPES and 37% of electricity generation. The importance of these energy sources contributes to Sweden's relatively low level of per capita CO_2 emissions compared to other IEA countries (further aided of course by the importance of nuclear power, which accounts for approximately 50% of electricity supply).

The recent direct and indirect measures that were introduced in Sweden to promote different renewable energies has helped raise their use, although the impact of indirect measures, such as the CO_2 tax introduced in 1991, is difficult to evaluate. The emphasis of short-term direct measures (e.g. subsidies and R&D) is on aiming to improve the economic attractiveness and competitiveness of proven renewable energy technologies. Longer-term R&D aims at development of new renewable energy technologies.

POLICIES

Economic and environmental issues are both important in Swedish energy policy, which places increasing emphasis on sustainable development. This is reflected in Sweden's pattern of renewable-intensive energy supply. Increased use of renewable energy sources are encouraged by subsidies, tax exemptions, guaranteed electricity purchases from small power producers, information programmes, and R&D and demonstration projects.

The Parliamentary decision on energy policy in June 1997 Toward a Sustainable Energy Supply, included a strategy for reducing the energy sector's impact on climate. The strategy is based on the view that successful international co-operation requires an equitable distribution of commitments and mitigation costs, and that national circumstances (e.g. mitigation measures already undertaken) should be taken into account when determining environmental commitments. Under the EU's burden sharing agreement, the EU's emission reduction commitment agreed to in Kyoto translates into a 4% increase in greenhouse gas emissions in Sweden for 2008-2012 compared to 1990 levels.

In January 1998, the Swedish National Energy Administration was set up. This body has the main responsibility for running the newly introduced energy programmes described below.

A seven-year programme aiming at an ecologically sustainable energy system was initiated in January 1998. Total programme funds of 5.28 billion SEK[60] are available over the seven-year period, including 2.73 billion SEK on energy research in Sweden. This reverses the previous downward trend in government R&D expenditure. An additional 1.61 billion SEK is dedicated to the support of commercial electricity production from renewables.

Another government programme set up to encourage increased use of renewables has been running since July 1997. This programme supports renewable energy investments in order to encourage increased production of renewable electricity, particularly from biomass and wind. Grants available since July 1997 are:

■ 25% for investments in CHP plants based on biomass (up to 3000 SEK/kWe), with a 5-year budget of 450 M SEK;

■ 15% for wind turbines over 200 kW, with a 5-year budget of 300 M SEK;

■ 15% for environmentally friendly, small-scale (<1.5 MW) hydro plants, with a 5-year budget of 150 M SEK.

This compares to similar grants for CHP and 35% grants for wind turbines > 60 kW under the previous investment support programme, initiated in July 1991. In addition to the 1997 investment support programme, the government set up a 5-year technology procurement programme for renewable electricity production from January 1998. Total funds for the procurement programme are 100 M SEK.

Holders of so-called supply concessions (i.e. electricity utilities) are obliged by the *Electric Law* to buy power from small power producers (< 1500 kW). The price for this electricity should be "fair", and this is defined as being the average sales revenues of the distributor, with deduction for reasonable costs of administration and a profit margin. In addition, small generators will obtain discounts on the network tariff and are exempt from the NOx levy, which only applies to plants with a production of over 25 GWh/year. The Government has provided for favourable buy-back rates for renewable-generated electricity from small producers further augmented by an "environmental bonus" (equal to the excise tax on

---

60   On average in 1997, I US $ = 7.635 SEK

Table 1
Energy and Environmental Tax rates from 1 January 1998 (SEK/unit)

	Residential/commercial rate	Industrial rate
Gas-oil (per m³)	1801	529
Heavy fuel oil (per m³)	1909	637
Coal (per ton)	1386	610
LPG (per ton)	1257	356
Natural gas (per 1000 m³)	1033	396
Petrol, leaded and unleaded (per litre)	4.47	-
Diesel (per litre)	2.67	-

electricity) paid for by the Government, when generation is by wind power. The exact amount of this tax varies within Sweden. "Green power" is also an available option for some consumers, e.g. wind power from Vattenfall.

Biofuels are not subject to energy taxes, i.e. they are exempt from the energy tax, CO_2 tax and sulphur tax. As a result of this exemption, a number of coal-fired CHP and district heating plants have changed to firing solid biomass. Since 1993, industrial users pay no energy tax and the carbon tax is applied at the lower rate of the rate applicable to other sectors.

Although fuel used in electricity production is exempt from tax, the carbon and energy taxes have helped to change the economics of new power generation options, making coal-fired district heating plants more expensive than any other option. The Energy Commission's report, released in December 1995, lists the following costs for electricity generation options (inclusive of taxes): see figure 1. The figures are for new CHP plants and for wind turbines in favourable locations.

The National Energy Administration, in co-operation with local authorities, is at present studying the possibilities of establishing areas of "national interest of wind power" in especially windy areas. Local authorities should, if areas are established, rate the interest of wind power highly when considering planning applications in these areas, and so give planning applications for wind power a greater chance of succeeding. Solar power is used commercially in Sweden in certain specific applications where the alternative cost to connect the application or the user to the grid is higher, e.g. remote houses, road section posts, beacons etc.

R&D is a fundamental component of Sweden's energy policy programmes. The downward trend of government R&D was reversed in 1998 with the implementation of the seven-year programme aiming at sustainable development.

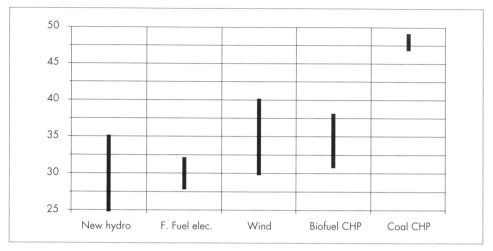

Figure 1
Cost of new electricity generation options in Sweden (öre/kWh)

The funds for this programme include 2.31 billion SEK over seven years for energy research, 210 M SEK for socio-economic studies, 210 M SEK on ethanol production and 870 M SEK on the Energy Technology Fund. Total R&D expenditure will be greater as funds e.g. from the Energy Technology Fund have to be complemented with funds from industry or other energy users.

The main purpose of the R&D programme is to reduce renewable energy's costs to make them a more economically viable alternative to fossil fuels. Short and medium-term measures are directed primarily towards an increased use of biomass, but also wind, hydro and heat storage. Longer-term priorities include fuel cells and solar heating, and focus on sustainable exploitation of renewable energy sources and increased efficiencies in the conversion technologies for heat and electricity production. These include many biomass-related projects, including one aiming to develop a new production process for ethanol based on cellulosic raw materials. Wind energy research was allocated 8 M SEK in FY 1995 by the Wind Power Consortium (VKK), created in 1994. An R&D programme on thin-layer PV cells is also supported with 15 M SEK for a 3-year period.

STATUS OF RESOURCE EXPLOITATION

The total importance of renewable energy (including hydro) in TPES has climbed from its 1970 value of 17% to a high of almost 27% in 1993. In 1996, hydro and non-hydro renewables accounted for 23% of TPES, as the very low rainfall in that year resulted in a lower-than-average utilisation of hydro. Non-hydro renewable energy use has grown rapidly from 5.5 Mtoe in 1990 to 7.6 Mtoe in 1996, of which biomass accounted for 7.3 Mtoe. The majority of electricity generation (89% in 1996) is met by nuclear and hydropower, although small amounts of electricity are also generated from waste, wind and solar power.

Biomass

Biomass accounted for 7.2 Mtoe (13.8%) of Sweden's TPES in 1996, and has grown in importance steadily since 1990, when use stood at 5.2 Mtoe. The majority of this is black liquor (3.1 Mtoe in 1996) and vegetal waste (3.0 Mtoe), with the remainder made up mainly from wood use. Biomass is projected to have a total potential of 146-169 TWh by the year 2000, roughly two-thirds from wood and derived products, with the remainder from energy crops (agricultural fuels). However, this potential far exceeds the expected demand from the market.

The majority of biomass, 5.1 Mtoe, is used directly, with the remainder being used to generate electricity and heat. In 1996, around 4.1 Mtoe was used in industry, mainly in the paper, pulp and wood industries, and 1 Mtoe was used in the residential sector in the same year. The use of solid biomass in public CHP systems has increased significantly between 1990 and 1996, when it reached 13 TWh, largely due to the introduction of carbon taxes in 1991. An impressive, although less spectacular, jump of over 80% can be seen in the use of solid biomass for heat production over the same time period. Much of the biomass used in CHP/district heating is co-fired with coal and oil. The capacity of such cofiring units was 1240 MW at the end of 1996. The government estimate that biomass-based CHP can be expanded by approximately 0.75 TWh per year between 1998-2003.

Waste

The use of municipal waste, predominantly for heat generation in the public sector, more than doubled between 1980 and 1996, when waste use reached an estimated 373 ktoe. A small amount of electricity, varying between 80 and 130 GWh since 1990, is also generated from municipal waste. Waste-to-energy utilisation is expected to increase further due to decreased landfilling.

Wind

Wind capacity and generation has been growing steadily since the beginning of the decade and by the beginning of 1998 there were 336 turbines in operation with a capacity of 123 MW. Wind production was 144 GWh in 1996, and is expected to be 240 GWh (0.18% of total electricity) in 1998. This compares to an estimated 3-7 TWh on-shore potential (after considerations for e.g. the military, other national interests, nature preservation, environmental and institutional aspects have been taken into account) and an additional 20 TWh off-shore potential. The government aims to increase annual electricity production from land-based wind power by approximately 0.5 TWh over five years.

Solar

No production of solar electricity is reported to the IEA. However, national estimates for installed PV capacity were 1.04 MW in 1994, mainly in test sites and of which only 10 kW was grid-connected. Increased use of PV power systems is restricted to specific applications. Direct use of solar energy was reported at 1 ktoe (12 GWh) in 1996, and is expected to grow to 57 GWh in 1998.

Hydro

Hydropower generally produces around half of Sweden's electricity, the majority from public power stations. However, its importance slipped back somewhat in 1996 (to 51 TWh and 37% of electricity production) because of substantially lower rainfall. Of Sweden's theoretical hydro potential of 130 TWh, only 5 TWh of non-exploited economic potential remains (after development restrictions for environmental reasons have been taken into account). However, the public acceptance for building new hydropower in Sweden is extremely low, and considering this low acceptance the Swedish National Energy Administration estimates that potential will increase only 1 TWh from its 1996 value to 2005.

Geothermal, Passive solar, heatpumps

Heat pump uses from ambient heat (air, soil, rock and sewage water) increased from 1 ktoe in 1990 to 12 ktoe in 1996. Passive solar is estimated to provide 13 TWh of heat per year (1993-1998).

Table 2
Trends in renewable energy supply and use

	Unit	1990	1992	1995	1996	2000	1990-1996 (%)[1]	1996-2000 (%)[1]
Renewable TPES (excl. hydro)	**ktoe**	**5507**	**5952**	**7203**	**7641**	**8027**	**5.6%**	**1.2%**
Percentage of TPES	%	11.5	12.7	14.1	14.5	15.4		
Geothermal	ktoe	0	0	0	0	0	n.a.	n.a.
Solar, Wind, Wave, Tide	ktoe	1	3	9	13	34	71.0%	27.4%
Biomass and Wastes[2]	ktoe	5506	5950	7194	7628	7993	5.6%	1.2%
- Biomass	ktoe	5152	5584	6797	7255	n.a.	5.9%	n.a.
- Wastes	ktoe	355	365	397	373	n.a.	0.8%	n.a.
Renewable electricity generation (excl. hydro)	**GWh**	**1893**	**2083**	**2523**	**2936**	**4470**	**7.6%**	**11.1%**
Percentage of total generation	%	1.3	1.4	1.7	2.1	3.0		
Geothermal	GWh	0	0	0	0	0	n.a.	n.a.
Solar, Wind, Wave, Tide	GWh	6	31	99	144	400	69.8%	29.1%
Biomass and Wastes[2]	GWh	1887	2052	2424	2792	4070	6.7%	9.9%
- Biomass	GWh	1758	1968	2308	2697	n.a.	7.4%	n.a.
- Wastes	GWh	129	84	116	95	n.a.	-5.0%	n.a.
Renewable TFC (excl. hydro)	**ktoe**	**4635**	**4554**	**5071**	**5056**	**5617**	**1.5%**	**2.7%**
Percentage of TFC	%	14.4	13.7	14.4	14.0	15.2		
Geothermal	ktoe	0	0	0	0	0	n.a.	n.a.
Solar, Wind, Wave, Tide	ktoe	0	0	0	1	n.a.	n.a.	n.a.
Biomass and Wastes[2]	ktoe	4635	4554	5070	5056	5617	1.5%	2.7%
- Biomass	ktoe	4630	4550	5068	5055	n.a.	1.5%	n.a.
- Wastes	ktoe	5	4	2	1	n.a.	-23.5%	n.a.
Hydro TPES	ktoe	6235	6393	5857	4424	5549	-5.6%	5.8%
Hydro electricity generation	GWh	72503	74332	68102	51444	64530	-5.6%	5.8%
Percent of total generation	%	49.7	50.9	45.9	36.9	42.8		

Notes:

1. Annual Growth Rate

2. Including Animal Products and Gases from Biomass

SWITZERLAND

OVERVIEW OF RENEWABLE ENERGY POLICY

Non-hydro renewable energy accounted for 5.7% of Swiss TPES in 1996, significantly above the IEA average of 3.9%. New renewable sources of energy could make an increased contribution to Swiss energy supply, and the Government in its *Energy 2000 Action Plan* has adopted targets for the future use of hydro and non-hydro renewables. To meet these targets, the Government, in co-operation with the cantons, has initiated a number of policies aimed at overcoming short-term institutional and market barriers. These efforts include direct and indirect subsidies, increased information dissemination, consultancy, training and marketing centres, and R&D. In addition, there is widespread use of voluntary measures.

The renewable energies promoted in Switzerland are also diverse, encompassing those used for electricity and heat generation, as well as for direct use. However, penetration of renewables into the Swiss energy market is likely to be very fragile for some time to come and will require continued marketing and promotion to build on the successful R&D programmes of the past few years. This will be particularly important for markets where subsidies have been used to start commercialisation.

The evaluations undertaken by the Swiss government indicate that the targets set for 2000 may still be achievable. To date, direct measures to promote renewables (e.g. via subsidies) have had a greater impact on their increased use than indirect measures, such as voluntary agreements. However, although the funds available for renewable energy promotion were increased in 1997, demand for subsidies has outstripped the funds available. Additional funds may be required in order to continue the growth in renewable energy use.

POLICIES

Increasing environmental awareness, a moratorium on new nuclear power plants and the prospect of Switzerland becoming a net electricity importer instead of exporter has influenced Swiss energy policy developments over recent years. Switzerland's overall energy policy initiative is *Energy 2000*: this came into effect in 1991 and aims to stabilise total fossil fuel consumption and CO_2 emissions at 1990 levels by 2000. Increased use of renewable energy features prominently in different parts of the *Energy 2000* programme, and will continue to be promoted post-2000 as the government's recent energy policy dialogue that ended in June

1997 reaffirmed a longer-term commitment to energy efficiency and renewable energy.

Promotion of renewable energy is distributed among federal, canton and municipal administrations. Policy measures used emphasise voluntary measures, information dissemination, R&D measures, financial incentives and regulations (including guaranteed markets for renewable electricity). Specific targets are set for the use of some renewable energy applications in 2000 (Table 1). "Green power" is also available for some electricity users.

The federal *Energy 2000* programme rests on three main pillars: voluntary measures to promote energy efficiency and renewables, a favourable legal environment, and dialogue between parties. There are many sub-elements of the *Energy 2000* programme, some of which are used in the promotion of renewable energy. These include the Federal Energy Office's programmes on renewable energy and funds for pilot and demonstration plants. Other programmes initially used to promote renewable energies have been phased out. These include the "start" programme, which ran from 1992-1995 and targeted the promotion of a particular renewable energy application/technology over a short period of time (e.g. PV systems in school buildings). DIANE, which was mainly used to encourage wood energy and small scale hydropower, ran from 1992-1997. The *Decree on Efficient Energy Use,* which has run since 1991 and guarantees a market for

Table 1
Targets for Renewable Energy under the Energy 2000 Programme

Energy source	Target by 2000	Comment
Non-hydro renewables (excluding wastes)	Additional 0.5 percentage point of electricity generation (compared to 0.01% of total generation in 1990)	Electricity generation from non-hydro renewables has been growing, and stood at 0.3% (approximately 60%) of total generation in 1996.
Renewables	3% of total heat production	May be met on schedule. By end 1996, 46% of the target had been achieved (mainly from increased use of wood-based heat).
Hydropower	5% increase in generation compared to 1990	A 1997 evaluation* indicated that 75% of the target is likely to be met by 2000. 1996's drop in hydropower was caused by low rainfall.
PV	Install 1 PV system in each Swiss commune.	Unlikely to be met on schedule, as interest in PV is concentrated in some parts of the country.
Heat pumps	100,000 installations	Would require accelerated installation rate to be met on schedule.

* Source: *7th Annual report on the Energy 2000 Action Programme,* Swiss Federal Department of Transport and Energy, November 1997

renewable electricity other than hydro from plants > 1 MW will be stopped at the end of 1998 and replaced by the *Energy Law* (when this is passed by the upper legislative body of parliament).

The Swiss parliament adopted a decree on federal investment in April 1997 that made an additional 64 M SF available for investments in the energy field between 1997-1999. Demand has been extremely high, and all 64 M SF had been allocated by May 1998. The federal government also made 200 M SF available from 1997 for the refurbishment of public infrastructure (which can include energy infrastructure). Available subsidies are generally of the order of 10%, but new systems using renewable energy can be subsidised up to 20%. Many cantons have programmes to promote renewable energy systems, but cantonal funds and staffing for the promotion of such systems have been cut over the past few years.

Promotion of renewable energy in Switzerland is more diverse than in many IEA countries, with emphasis placed on increased use of wood and solar energy for heating purposes, increased use of heat pumps, and on increased renewable electricity generation. However, budgetary constraints have meant that available funds are lower than initially expected (57 M SF/year instead of the anticipated 170 M SF), which is resulting in increased emphasis being placed on indirect promotion of renewable energy via for example information programmes and voluntary agreements, and to a reduction in the financial incentives offered for renewable energy.

The Government has made significant efforts in marketing information on renewable energy sources through its network of local energy advisory centres. In 1994, 70 centres provided consultancy services, mostly free of charge, on the use of renewable energy sources and energy-efficient technology to end-users and local authorities.

The implementation of *Energy 2000* rests with 8 'marketing departments', one of which deals specifically with renewable energy. The work of this department is concentrated in three areas with an associated Action Network. Swissolar assembles five utility and private solar energy associations, and its main purpose is the promotion and marketing of solar energy technologies. The Swiss Wood Energy Association aim to increase wood's share of the heating market to 6% by 2000. There is also a Swiss Heat Pump Promotion Group which leads promotional efforts such as training, quality assurance and after-sales service. However, almost 10,000 heat pumps would need to be installed each year to 2000 if the Energy 2000 goal is to be met.

The Government expects 1 500 GWh of heat (approximately 50% of the "Energy 2000" target for heat from new renewable energy sources) to be produced through voluntary action by the Swiss cement industry, which has agreed to substitute waste-derived fuels for 75% of its current coal consumption.

Evaluation of renewable energy policies

The Swiss government is one of only a few IEA countries that has undertaken evaluations of renewable energy policies to determine the relative effectiveness of different policy measures. In addition, annual progress reports towards the *Energy 2000* goals are also published.

By 1997, 7651 installations had received 182 M SF subsidies under the *Energy 2000 Action Programme*. These installations had an installed capacity of 497 MWth and 10.1 MWe. This excludes any installations undertaken as a result of indirect (e.g. voluntary) measures. In 1996, the renewables component of *Energy 2000* was estimated to have reduced emissions by 105-115 kt CO_2 (primarily via subsidies and other direct measures).

Policies directed towards increased penetration of renewables in the domestic sector, i.e. increased use of heat-pumps, have resulted in a significant number being installed – albeit at a lower rate at present than hoped for. The most important factor in influencing people's choice to install a heat pump was found to be the availability of adequate information. Surprisingly, 85% of people questioned stated they would have acted in the same way even if no subsidies were in place. Heat pump subsidies for indviduals were therefore found to be of limited use. Evaluation of policies designed to encourage increased use of solar collectors in dwellings have indicated that some consumers are willing to pay "two or three" times the price for solar energy than for normal energy. Funds are therefore best directed towards increased availability of information, since a large proportion of people questionned (40% in the solar collector evaluation study) were unaware of the promotional programme.

Subsidies were, however, more important to co-operative and private organisations: the former accepted additional costs for environmental purposes more readily than private business, who wanted investments with payback periods of less than 10 years. Recent recommendations of the Federal Energy Office suggest maintaining or increasing the level of subsidy for these target groups, (especially in the area of wood for heating).

The latest (1995) figures available for federal energy R&D expenditure indicate a high level of expenditure on renewables $43m: this accounts for around a quarter of overall federal energy R&D expenditure (one of the highest in the IEA). The majority of funds are allocated to solar applications (mainly solar heating and cooling), followed by biomass.

Green electricity is available in some parts of the country. In the Bernese Electric Utility customers can opt to buy wind electricity at generation cost price: i.e. the price paid for wind electricity would be higher than that for electricity from other sources. PV electricity is also sold at 1.1 SF/kWh (compared to a normal household electricity price of 0.2 SF/kWh).

STATUS OF RESOURCE EXPLOITATION

Non-hydro renewable energy contributed 1.5 Mtoe to Swiss TPES in 1996, up 49% from the level in 1990. Hydropower contributed an additional 9.5% of energy supply and generated 51% of electricity in 1996. The major non-hydro renewable energy used is wastes and biomass. These are used both for electricity and heat generation, and in direct use applications e.g. in industry. There is also some use of solar heat and electricity, but wind power is tiny.

Biomass

Use of biomass (mainly wood) has been growing steadily during the 1990s, and contributed 554 ktoe to TPES in 1996. The majority of this was used directly in the residential sector, although also in industry and the commercial sector. Small amounts of wood, landfill gas and sewage sludge were also used to generate electricity (142 GWh in 1996).

Waste

Municipal and industrial wastes together contributed 847 ktoe to Switzerland's energy supply in 1996, up from 554 ktoe in 1990. Approximately three-quarters is used for electricity and heat generation: 1030 GWh in 1996, and 12340 TJ (almost a doubling of their values in 1990). The remainder is used directly in industry (predominantly the chemical industry). There is only limited potential to increase energy extraction from wastes, and is unlikely to increase significantly beyond 2000.

Wind

General conditions are viewed by the Government as unfavourable to wind generation. Consequently, capacity and generation are small: 2MW and 1 GWh in 1996. The government does not expect this to increase substantially by 2000, although projects wind generation to take off after the turn of the century.

Hydro

Almost 12 GW of installed hydro capacity in 1996 generated 28403 GWh of electricity. This was substantially lower than generation between 1994-1996 due to lower rainfall. Around 800 small scale (<300 kW) hydro installations are in use generating around 270 GWh from 60 MW. The theoretical potential for hydropower development is much higher, although costs are rising with the increase in annual water royalties paid to cantons and municipalities and as new sites close to demand centres become increasingly hard to find. The Mauvoisin electricity utility presented their December 1995 decision to indefinitely postpone a major (550 MW) capacity upgrade of the Mauvoisin hydro project as an end to all major Swiss hydro projects in the foreseeable future. The upgrade was postponed for a number of reasons including the economic recession in Europe, falling electricity prices and increasing water royalties. Small-scale hydro plants (< 1 MW) are exempted from the increase in water royalties, and systems between 1-2 MW subject to a reduced rate only, in order to encourage increased use. Given normal precipitation conditions, hydro generation could in theory be raised a further 570 GWh by 2000.

Solar

Active solar thermal installations accounted for the majority of solar capacity installed in Switzerland and was estimated by the Administration at 14.5 MW in 1996. Solar heat used stood at 17 ktoe. In addition, there were grid-connected PV installations with a capacity of 7 MW that generated 7 GWh. However, data reported to the IEA for 1996 indicated a total solar capacity of 9 MW. This has been growing steadily since 1990, and growth is expected to continue past 2000 despite the high cost of electricity from PV systems.

Geothermal

No energy use reported to the IEA.

Heat pumps

By the end of 1994, around 43,000 heat pumps were installed in Switzerland, with the heat extracted standing at 65.5 ktoe. This had reached almost 50,000 by 1996. The number of heat pumps is set to increase as the *Energy 2000* target is for 100,000 units to be installed by 2000 (although this target is unlikely to be met on time).

Table 2
Trends in renewable energy supply and use

	Unit	1990	1992	1995	1996	2000	1990-1996 (%)[1]	1996-2000 (%)[1]
Renewable TPES (excl. hydro)	**ktoe**	**990**	**1061**	**1439**	**1473**	**1406**	**6.8%**	**-1.2%**
Percentage of TPES	%	4.0	4.2	5.7	5.7	5.7		
Geothermal	ktoe	0	0	0	0	0	n.a.	n.a.
Solar, Wind, Wave, Tide	ktoe	0	0	16	18	1	n.a.	-51.3%
Biomass and Wastes[2]	ktoe	990	1061	1423	1455	1405	7.5%	-0.3%
- Biomass	ktoe	436	472	595	608	n.a.	6.4%	n.a.
- Wastes	ktoe	554	588	828	847	n.a.	8.4%	n.a.
Renewable electricity generation (excl. hydro)	**GWh**	**559**	**616**	**1056**	**1180**	**609**	**13.6%**	**-10.4%**
Percentage of total generation	%	1.0	1.1	1.7	2.1	1.0		
Geothermal	GWh	0	0	0	0	0	n.a.	n.a.
Solar, Wind, Wave, Tide	GWh	0	2	6	8	9	n.a.	8.4%
Biomass and Wastes[2]	GWh	559	614	1050	1172	600	13.4%	-10.6%
- Biomass	GWh	8	9	144	142	n.a.	78.3%	n.a.
- Wastes	GWh	551	605	906	1030	n.a.	10.5%	n.a.
Renewable TFC (excl. hydro)	**ktoe**	**596**	**672**	**691**	**760**	**1010**	**3.0%**	**7.9**
Percentage of TFC	%	3.0	3.3	3.4	3.7	5.2		
Geothermal	ktoe	0	0	0	0	0	n.a.	n.a.
Solar, Wind, Wave, Tide	ktoe	0	0	15	17	n.a.	n.a.	n.a.
Biomass and Wastes[2]	ktoe	596	672	676	743	1010	2.5%	8.4%
- Biomass	ktoe	436	472	474	528	n.a.	1.7%	n.a.
- Wastes	ktoe	160	200	202	215	n.a.	4.7%	n.a.
Hydro TPES	ktoe	2562	2812	3025	2443	2933	3.4%	0.6%
Hydro electricity generation	GWh	29795	32700	35169	28403	34100	3.4%	0.6%
Percent of total generation	%	54.6	56.6	56.5	51.0	58.0		

Notes:

1. Annual Growth Rate

2. Including Animal Products and Gases from Biomass

TURKEY

OVERVIEW OF RENEWABLE ENERGY POLICY

The aims of Turkish energy policy are to satisfy energy demand consistent with economic, development and environmental objectives. Non-hydro renewable energy use accounted for 11.1% of TPES in 1996 – almost triple the IEA average – largely comprising the direct use of biomass. Turkey differs from most IEA countries, having a significantly lower per capita GDP and energy use. In addition, the rapid growth in energy and electricity demand means that electricity shortfalls are likely to occur in the short term. Turkey's renewable energy profile is also significantly different from the average: the majority of renewable energy supply is biomass which is used almost exclusively within the residential sector for heating. Very little "new" renewable energy is used.

Turkey has no national renewable energy policy, and there are few Government-backed incentives to promote renewable energy. Turkey is planning to develop the use of geothermal heat (projected to rise from under 62 ktoe in 1996 to almost 685 ktoe in 2000). However, the pace of total energy growth is likely to be more rapid than that of renewable energy development, and the percentage of renewables in total energy supply is likely to continue declining to 2000 and beyond.

The Ministry of Energy and Natural Resources (MENR) is preparing draft legislation which would allow certain renewable energy projects (mainly geothermal and wind, but also solar, wave, waste and landfill gas only) to be built and operated by the private sector, and provide incentives for such systems. This legislation would also set the buy-back rates for renewable electricity.

POLICIES

Turkey's energy policy priorities are to ensure a sufficient, reliable and economic energy supply and to meet current and anticipated rapid expansion in energy demand. Energy supply and consumption grew at around 4% p.a. between 1990 and 1996, and almost 8 % for electricity consumption over the same time period. Rapid growth rates are expected to continue in the near term, with TPES set to increase by 8% p.a. between 1996-2000, compared to an IEA average of 1.1%, and electricity output projected to increase 9.1% p.a., compared to an IEA average of 1.6% p.a.. Development of renewable energy sources is an integral part of efforts to increase total energy production, and to more fully exploit indigenous energy resources.

Both the Ministry of the Environment and (indirectly) the Ministry of Energy and Natural Resources (MENR) are involved in renewable energy promotion. The Ministry of Environment promotes the development of geothermal heat and other environmentally-friendly investments via low-interest loans on up to 45% of the capital cost. MENR acts via the Electrical Power Resources Survey and Development Administration (EIE) , which carries out a number of different tasks including studies on solar and wind power as well as R&D.

Turkey is not a signatory of the FCCC and has therefore no greenhouse gas-related commitments[61]: Turkey is also not included in the list of countries with greenhouse gas commitments under the Kyoto Protocol. However, a project sponsored by the World Bank, 'Turkey National Environmental Strategy and Action Plan' includes measures aimed to promote development of renewable energy technologies and emphasises in particular the efficiency of renewables as a way of minimising CO_2 emissions.

MENR has announced a target for wind energy of 2% of total installed capacity by 2005. There are no other national renewables targets, although one municipality (Greater Ankara) has a local target of providing 10% of its energy needs from renewables (mainly solar heat, PV and wind-generated electricity) by 2005. Within the context of this local plan, studies are being undertaken on passive solar buildings with the intention of incorporating successful designs into housing legislation. There is some municipal support in the areas of hydro and geothermal heat. Private sector involvement in renewable energy promotion exists, predominantly in areas for wind energy and small scale solar projects.

To meet forecast increases in electricity demand, the government has a generating capacity expansion plan to increase generating capacity from 20 GW in 1996 to 28 GW in 2000 and 65 GW in 2010. The share of fossil fuels is likely to increase and that of hydropower is expected to decline slightly to 37% of electricity production by 2010. Independent electricity producers, including those using renewable energy sources, are given a power purchase guarantee by the Turkish Electricity Generation and Transmission Corporation (TEAŞ), who report to the MENR. The purchase price for independent generation was increased by 21% in 1998 – partly because increased independent generation will be needed to fill Turkey's projected short-term power deficit (estimated to reach 3 TWh in 1998). The Government has planned (and implemented) a forest management programme to reduce uncontrolled woodcutting.

Total expenditure of renewable energy R&D for 1996 was reported to the IEA as 0.15 M USD, representing 4.6% of total energy R&D. The main resources being supported are solar, geothermal, and wind, all of which have a large and mainly untapped potential in Turkey and so could help to meet its rapidly growing energy

---

61 Turkey may ratify the UNFCCC if this is amended to exclude Turkey from the list of countries with greenhouse gas and financial commitments.

demand. However, other R&D on the demonstration of advanced biofuels technology, such as electricity generation from biomass and liquid biofuel development are also underway.

STATUS OF RESOURCE EXPLOITATION

The share of renewable energy in Turkey's TPES in 1996 was 11.1%: almost triple the IEA average, with the majority of total renewable energy supply from biomass and animal products, mostly wood, and some geothermal energy. The first wind turbines started operating in early 1998. In addition, hydro electricity provided 43% of country's electricity production. The relative importance of renewables in TPES continues to decline as domestic coal production and natural gas imports rise steadily to satisfy rising electricity demand. However, there is significant room for expanding renewable energy use in Turkey, which has a large resource base of biomass, geothermal, solar and wind energy as well as hydro.

Biomass

Utilisation of biomass is expected to increase very slightly by 2010 to 7.2 Mtoe. However, the share of renewable energy use made up from direct use of biomass will drop substantially, as other renewables (most notably hydro – see below) are set to expand greatly.

Waste

No utilisation of waste for energy purposes is reported. However, four autoproducer waste-to-energy plants with a combined capacity of 86 MW are being planned.

Wind

Turkey's first wind farm was commissioned in February 1998, and has a capacity of 1.5 MW. Capacity is likely to grow rapidly, as plans have been submitted for just under a further 600 MW of independent facilities (of which 57 MW is at an advanced stage in negotiations). The majority of proposed projects are located in the Çeşme-Izmir and Çanakkale regions.

EIE carries out wind measurements at various locations to evaluate wind energy potential over the country, and have started to compile a Wind Energy Atlas (in conjunction with other organisations). Approval of independent wind energy projects requires at least a 6-month history of wind measurements.

Solar

The direct use of solar heat is small, but growing rapidly: Turkey estimate that this contributed 80 ktoe in 1996, more than four times its level in 1990, with 2.5 million m² of solar thermal collectors in place. However, this is not reflected in statistics reported to the IEA. No solar power is currently produced.

Hydro

Turkey has an economic hydropower potential estimated at 124 TWh, of which almost two-thirds has not yet been exploited. Hydro's share of electricity output in 1996 – a wet year – reached 43%, when it generated almost 41 TWh (almost double its 1990 level). Increased generation is expected to continue, with generation rising to almost 44 TWh in 2000 and 85 TWh in 2010. Capacity is set to more than double over the next 15 years, with its 1996 level of 9.9 GW rising to 15.7 GW in 2000 and 24.5 GW in 2010. The vast majority of hydro capacity is large hydro: systems smaller than 10 MW account for less than 1% of total capacity in 1996.

However, the growth rate in electricity demand is set to outpace that of hydro development, and the share of hydro power in electricity generation is expected to drop to 32% of the total by 2000. Hydropower accounts for over 99% of renewable electricity generation – a share that is not likely to drop in the foreseeable future.

Geothermal

Geothermal's contribution to TPES was 162 ktoe in 1996: including 90 ktoe direct use of geothermal heat. The remainder was used to generate 84 GWh of electricity. Official forecasts predict significant increases in direct use of geothermal energy for heating to 4567 ktoe in 2010, as the government is working on legislation aiming to increase the use of geothermal energy.

Table 1
Trends in renewable energy supply and use

	Unit	1990	1992	1995	1996	2000	1990-1996 (%)[1]	1996-2000 (%)[1]
Renewable TPES (excl. hydro)	**ktoe**	**7311**	**7328**	**7255**	**7285**	**7769**	-0.1%	**1.6%**
Percentage of TPES	%	13.9	13.3	11.7	11.1	8.5		
Geothermal	ktoe	85	90	138	162	685	11.4%	43.4%
Solar, Wind, Wave, Tide	ktoe	21	32	52	80	121	25.0%	10.9%
Biomass and Wastes[2]	ktoe	7205	7206	7065	7043	6963	-0.4%	-0.3%
- Biomass	ktoe	7205	7206	7065	7043	n.a.	-0.4%	n.a.
- Wastes	ktoe	0	0	0	0	n.a.	n.a.	n.a.
Renewable electricity generation (excl. hydro)	**GWh**	**80**	**117**	**308**	**260**	n.a.	**21.7%**	**n.a.**
Percentage of total generation	%	0.1	0.2	0.4	0.3	n.a.		
Geothermal	GWh	80	70	86	84	90	0.8%	1.7%
Solar, Wind, Wave, Tide	GWh	0	0	0	0	0	n.a.	n.a.
Biomass and Wastes[2]	GWh	0	47	222	176	n.a.	n.a.	n.a.
- Biomass	GWh	0	47	222	176	n.a.	n.a.	n.a.
- Wastes	GWh	0	0	0	0	n.a.	n.a.	n.a.
Renewable TFC (excl. hydro)	**ktoe**	**7242**	**7238**	**6904**	**7011**	**7692**	-0.5%	**2.3%**
Percentage of TFC	%	18.0	17.1	14.5	14.1	10.9		
Geothermal	ktoe	16	30	64	90	608	33.3%	61.2%
Solar, Wind, Wave, Tide	ktoe	21	32	52	80	121	25.0%	10.9%
Biomass and Wastes[2]	ktoe	7205	7176	6788	6841	6963	-0..9%	0.4%
- Biomass	ktoe	7205	7176	6788	6841	6963	-0.9%	0.4%
- Wastes	ktoe	0	0	0	0	0	n.a.	n.a.
Hydro TPES	ktoe	1991	2285	3057	3481	3762	9.8%	2.0%
Hydro electricity generation	GWh	23148	26568	35541	40475	43750	9.8%	2.0%
Percent of total generation	%	40.2	39.5	41.2	42.7	32.6		

Notes:

1. Annual Growth Rate

2. Including Animal Products and Gases from Biomass

UNITED KINGDOM

OVERVIEW OF RENEWABLE ENERGY POLICY

The *Non-Fossil Fuel Obligation* (NFFO) scheme, which provides output subsidies for renewable electricity from selected sources, and a complementary R&D programme, are the main measures in place to develop renewables in the UK. Other measures aimed at strengthening institutional support (provision of information, removal of market barriers and support for renewable industries) are also being actively pursued. The new UK Government is currently reviewing renewable energy policy[62] to see what would be necessary and practicable to achieve 10% of UK's electricity needs from renewables by the year 2010 and how renewables can make an effective contribution to meeting requirements for future greenhouse gas reductions commitments. The previous government had a shorter, non-binding, target of 1500 MW of installed renewable electricity generating capacity by 2000. Promotion of renewables is aimed at expanding an already diverse range of indigenous energy resources including fossil fuels and nuclear and hydro power and contributes to wider industrial and environmental objectives.

The 'fossil fuel levy' and the NFFO subsidy instruments are a combination of direct subsidy and market-oriented approach. Explicit targets for particular technologies are not adopted, although Government indirectly influences the mix of different technologies by selecting which projects benefit from NFFO. Emphasis is on the development of electricity rather than heat output from renewable sources.

The UK Government has estimated the accessible renewable resource (within certain cost limits) at over three times the UK's total electricity generation. Non-hydro renewable generation represents 1.7% of total generation in 1996. An additional 1 % come from hydro sources. If the UK Government adopts a target of 10% of UK electricity generation from renewable sources, this will require significant new investment in renewable energy plants. Current measures have clearly stimulated the development of wind, waste, landfall gas and biomass technologies resulting in 506 MW of additional installed capacity over 8 years.

POLICIES

The UK is an energy-rich country, with indigenous reserves of coal, oil and natural gas and abundant sources of renewable energy. Ensuring diverse, secure and

---

62 UK Energy Policy is described and analysed in detail in the IEA's forthcoming *Energy Policies of the United Kingdom, 1998 Review.*

sustainable energy supplies; encouraging an internationally competitive domestic renewable industry; and reducing emissions of polluting gases including CO_2 are important features of the policy context surrounding the UK Government's promotion of renewable energy. The UK Government is seeking to increase the use of renewable energy resources by giving them the opportunity to compete equitably in a self sustaining market.

In June 1997, the new UK Government announced its review of renewable energy policy to see what would be necessary and practicable to achieve 10% of UK's electricity needs from renewables by the year 2010 and how renewables can make an effective contribution to meeting requirements for future greenhouse gas reductions commitments. The outcome of the review will be published during 1998. Proposals for a fifth *Non Fossil Fuel Obligation* (NFFO-5) Order were also announced in November 1997 with a view to making the order in late 1998.

Measures in the following five complementary areas are in place to achieve renewable goals:

■ stimulating market conditions for renewable electricity production via the *Non-Fossil Fuel Obligation* NFFO (for a limited period only);

■ continuing R&D in promising areas;

■ ensuring an adequate flow of information on renewable energy;

■ removing market barriers that inhibit the uptake of renewable energy; and

■ encouraging internationally-competitive industries to develop.

The UK Department of Trade and Industry is responsible for administering policies and programmes to promote renewables. Local government planning authorities and the Environment Agency are responsible for some regulatory aspects of renewables.

The development of renewables is also affected by environmental legislation and regulations, planning legislation and regulations and agricultural policy and support initiatives. Other areas of policy having indirect effects are usually related to the above three in some way (e.g. the application of the UK building regulations) or are not thought to influence developments greatly.

The promotion of renewable energy in the UK has as one of its aims reduction in emissions of pollutants, including gases that may contribute to climate change. The UK Government has calculated that by 2000, renewables could contribute up to 2 million tonnes of carbon savings annually (7.3 mt CO_2), 100,000 tonnes of SOx, and 30,000 tonnes of NOx.

The NFFO is an output subsidy which guarantees a market for certain pre-competitive renewable electricity-generation technologies for a number of years and is the main thrust of Government support for renewables. NFFO operates through the *Electricity Act* (1989) which empowers the Secretary of State to make orders requiring the public electricity suppliers to secure specified amounts of renewable energy generation capacity from specified renewable energy sources. The public electricity suppliers meet these obligations by contracting with renewable energy generators at prices which provide a guaranteed market. Renewable energy suppliers bid in a competitive process to supply electricity at the lowest bid price. The regional electricity companies (RECs), who are responsible for distributing electricity to consumers, are required to ensure the distribution of a certain amount of renewable generated electricity. The RECs collaborate collectively through a Non-Fossil Purchasing Agency to meet these obligations, and sign contracts with renewables-based generators for their electricity. The integral role played by the RECs in the NFFO contractual process has helped focus their attention on renewable energy, in some cases to the extent that the RECs have become active in supporting renewable energy bids under NFFO. This has resulted in increased numbers of studies and analyses for schemes although much of this work remains commercially confidential.

The NFFO originated in 1989 as a mechanism for protecting nuclear power during the privatisation of the electricity supply industry. The measure was extended to include renewables in 1990. Since 1990 there have been four renewable Orders under the NFFO (referred to as NFFO-1, 2 , 3and 4 respectively) in England and Wales, two Scottish Renewables Orders (SRO) and two Orders in Northern Ireland (NI-NFFO). Further Orders are planned for all regions.

NFFO orders are financed through the 'fossil fuel levy' on electricity bills which originally ran to December 1998 for NFFO-1 and NFFO-2. Following agreement from the European Commission, the levy for renewables has been separated from that of nuclear power and for NFFO-3, 4 and 5 is not constrained by the 1998 deadline. The contractual timescale for NFFO now runs for 15 years within a 20 year time span, a period related in part to plant lifetime. NFFO support for 1997/98 was around 120M GBP.

NFFOs 1, 2, 3 and 4, together with the Orders in Scotland and Northern Ireland, have brought 506 MW of new renewable energy generating capacity on stream (as at December 1997). The capacity of bids approved under the NFFO orders to date is significantly higher than that actually constructed (Table 1) due to delays in construction or failure to obtain planning permission. A summary of the successful bids in NFFO-4 is shown in table 2.

The price paid for electricity from the different renewables has again declined (as illustrated in the above table) between NFFO-3 in 1994 and NFFO-4 in 1997. The NFFO-3 prices were also lower than the "strike prices" paid under NFFO-2. This increased renewables capacity and generation may help the government achieve some of its aims regarding energy security and diversity, and also contributes to lowering energy-related greenhouse gas emissions. However, whether this plank of the UK's renewable energy policy will help domestic industry is uncertain: wind power has benefitted from 70 M GPB Department of Trade and Industry (DTI) and

Table 1
Renewable Orders in the United Kingdom

	Number of projects	Approved capacity (MW)	Installed capacity (end 1997, MW)
NFFO-1 (1990)	75	152	144.5
NFFO-2 (1992)	122	472	181.5
NFFO-3 (1994)	141	627	138.1
NFFO-4 (1997)	195	843	0.7
SRO-1 (1994)	30	77	1.6
SRO-2 (1997)	26	114	27.1
NI-NFFO-1 (1994)	20	16	14.6
NI-NFFO-2 (1996)	10	16	0.1

Table 2
Summary of NFFO-4 and comparisons with NFFO-3 and NFFO-2

Technology Band	Contracted Capacity (MW dnc*)	NFFO-4 Weighted Average Price (p/kWh)	NFFO-3 Weighted Average Price (p/kWh)	Price paid in NFFO-2
Wind				11.0
> 0.768 MW dnc	330.36	3.53	4.32	
< 0.768 MW dnc	10.33	4.57	5.29	
Hydro	13.22	4.25	4.46	6.0
Landfill Gas	173.68	3.01	3.76	5.7
Waste-fired CHP	115.29	3.23	3.84	6.55 "waste" 5.9 "sewage gas"
Waste-fired FBC	125.93	2.75		
Biomass gasification	67.33	5.51	8.65	
Anaerobic digestion of agricultural wastes	6.58	5.17		
Other			5.07	5.9 "other"

* = declared net capacity

Department of Energy research grants to the end of 1996/97, and has 55 commissioned projects under the first three NFFOs, the first Scottish Renewables Order (SRO) and the first Order in Northern Ireland (NI-NFFO). Despite this (and the presence of a British turbine manufacturer) over 80% of turbines are imported.

The aims of R&D are to provide an assessment of technology options and potential; stimulate technology development and cost reduction with industry; monitor developments under the NFFO; remove inappropriate market barriers; and increase technology transfer. Resources are directed towards a number of 'key technologies' including hydro, landfill gas, municipal and industrial waste, wind, solar, fuel cells, energy crops and agricultural and forestry wastes. The previous government announced in 1994 that state funding for tidal, wave and geothermal energy would be halted at the end of their commitments in these areas because of the limited commercial potential in these areas, but this is being reviewed by the new Government.

Government R&D expenditure for FY 1996/97 declined by around 20% compared with 1995/96 to 14.7 M GBP. The R&D budget for PVs has more than doubled in recent years to reach 1 M GBP in 1998/99. From late 1994 onwards the UK R&D programme has seen the creation of a task force to assist the transfer of UK new and renewable energy expertise to overseas markets, and now employs two dedicated Renewable Energy Trade Promoters.

Short-rotation coppicing and other aspects of biomass exploitation are believed to be a promising area, and over six years from FY 1990/91 to 1996/97, this R&D budget has increased from about 0.9 M GBP to 2.7 M GBP per year. Activities within the UK's biomass R&D programme area come under three main headings including: agricultural and forestry residues; crops; and advanced conversion technologies. The intentions behind the programme area's activities are to drive down fuel production and utilisation costs, increase the efficiency of fuel production and use, and provide support to domestic industry to diversify into the biomass to energy market. Coppice crops are eligible for financial support under the Set-Aside and Forestry Authority Woodland Grants Scheme. The level of grant available depends on whether the land is designated as eligible for set-aside payments. For set aside land 400 GBP/ha/yr (up to a maximum of 1250 ha/yr) is available and 600 GBP/ha/yr (up to a maximum of 1000 ha/yr) is available for non-set aside land.

Information and advice on passive solar design of buildings is available on a regional basis, and the government estimates that this information could contribute to annual energy savings of 50M GBP/year by 2000. The aims of the UK passive solar design programme are to encourage the uptake of passive solar design, develop internationally competitive industries for the domestic and export markets, and to quantify the environmental improvements and disbenefits associated with passive solar design.

Government policies on waste affect the proportion of landfill gas that is produced by recovery schemes. Increasing the utilisation of methane from landfills has the double dividend of reducing emissions of a potent greenhouse gas while diverting energy demand from conventional sources. This is encouraged via the NFFO discussed above. Alternatively, the double dividend could result from incinerating waste that would otherwise have been landfilled. At the end of December 1997, 82 landfill gas plants had been commissioned under the NFFO and related schemes, comprising 151 MW.

The UK Department of Environment, Transport and Regions (DETR) has issued planning guidance for England on renewable energy (PPG22). Similar guidance has been issued by Regional Departments in Scotland, Wales and Northern Ireland.

The DETR issued a consultation draft Waste Strategy in January 1995 which elucidated three key principles for waste management: to reduce the amount of waste that society produces; to make best use of the waste that is produced; and to choose waste management practices which minimise the risks of immediate and future environmental pollution and harm to human health. These principles set a waste hierarchy of reduction, reuse, recovery and disposal.

The UK *Environmental Protection Act* (1990) set the framework for an improvement in landfill standards by instituting a system of landfill regulation and licensing to be administered locally and leading to greater scrutiny and control of all land filling operations. A levy on landfill was introduced in 1996. Even after the imposition of such a levy, landfill will remain the least cost option for many areas in UK.

Evaluations of the renewable energy programme are carried out at intervals by the UK DTI. The most recent of these reviews is summarised in Energy Paper No. 62 (1994).

STATUS OF RESOURCE EXPLOITATION

Fossil fuels provide the majority of energy produced and consumed in the UK, although nuclear power generates a significant proportion of electricity. Non-hydro renewable energy resources account for less than 1% of TPES, predominantly from biomass and wastes. Electricity generation from non-hydro renewables and waste was 5.5 TWh in 1996. This should increase considerably as more projects under NFFO-3 and NFFO-4 are commissioned.

Energy Paper 62 shows that, for electricity producing renewables, the maximum practicable resource at 2005 (assuming an 8% discount rate) is dominated by onshore wind and energy crops for costs less than 10p/kWh, with smaller but significant contributions from municipal and industrial waste combustion, landfill gas, conventional forestry and agricultural wastes and hydro.

Government assessments of the "accessible resource" for electricity production from renewable energy (defined as technologies capable of producing electricity for less than 10p/kWh (1992) are at 1 100 TWh per year, or over three times the

Table 3
Grid Connected Renewable Energy technologies in UK at December 1997

	Number of installations	Capacity (MW DNC)
Hydro	46	32.023
Landfill gas	82	151.088
Mun. & Ind. Waste combustion	9	132.83
Sewage gas	25	32.839
Other combustion	4	25.58
Wind	55	132.288
Total	221	506.648

Source: Communication from UK DTI

UK's total electricity generation in 1993. This estimate will increase with decreasing costs of renewable electricity, but has not yet been updated.

The current level of grid connection (as at December 1997) for different renewable energy technologies is shown in Table 3.

Biomass

Combustion of wood and straw generated an estimated 270 ktoe of useful heat per year since 1992 – over half of total renewables heat production. However, electricity generation from biomass (as distinct from landfill or sewage gas) had limited success under NFFO-3, with priority being given to three gasified biomass electricity generation projects totalling 19 MW. Conventional steam-cycle biomass boiler technology was not supported. Recently a contract was signed to develop an 8 MW power station fuelled by short rotation coppice, to be sited at Eggborough in North Yorkshire. Thirty five landfill gas generation sites with a combined capacity of 70.1 MW were generating electricity under NFFO-3 and 4 by December 1997. Sewage gas projects are now no longer being supported under the NFFO.

Waste

Municipal and industrial waste formed the largest category of successful renewable projects in both NFFO-3 and 4. The average price paid for municipal and industrial waste combustion schemes for current generation under NFFO-4 is less than 3.5p/kWh (see Table 2).

Wind

The UK's largest accessible renewable energy source is wind power: the accessible resource has been estimated at 340TWh/y onshore and 380TWh/y offshore. While all of this technical potential is unlikely to be economically feasible, notably the offshore resource, the 'practicable resource' for (onshore) wind power of 55TWh/year (as estimated by the DTI) represents a sixth of the UK's total electricity generation in 1993. Electricity actually generated from wind power in the whole of the UK in 1996 was 0.49 TWh, and was estimated at 0.66 TWh in 1997. This has been increasing steadily throughout the 1990s, due to the construction of wind electricity systems under the NFFO. Estimated capacity at end 1997 was 135.4 MW, double its value of 1994. Both capacity and generation are likely to increase further in the short-term as successful wind projects from NFFO-3 and 4 come on line, and in the longer term (assuming that wind systems continue to account for a large percentage of successful bids under NFFO-5).

Solar

The UK's PV installed capacity was estimated at 300 kW by the end of 1996/7. Support for PV through the NFFO is unlikely in the short term although prices per peak watt are expected to continue to fall into the next century. Capacity and generation are therefore unlikely to grow significantly in the short to medium term.

Hydro

Hydropower produced over 3.3 TWh in 1996, mainly from large-scale plants in Scotland. The estimated UK accessible resource for new small-scale hydro power is 3.9 TWh/yr of which 3.4 TWh/yr is within Scotland. The resource likely to be commercially attractive is however much smaller than this. Generation from small-scale plants in the UK was estimated at 159 GWh in 1997.

Tidal projects in the UK have been studied extensively. The technology is well understood which in turn suggests that there is little further scope for cost reductions. With this in mind, the economics of electricity generation from tidal energy in the UK do not look promising under the future scenarios envisaged.

The UK investigated more than 300 design concepts for wave energy devices over the period 1974 to 1983. Eight of these concepts were developed and tested as full-scale electrical generating systems, both in experimental facilities and in the sea. More recently a review of wave energy (1992) concluded that the main devices assessed were unlikely to generate electricity competitively in the short to medium term. The UK shoreline wave energy resource may become economic

under some future scenarios but the estimated accessible resource is no more than
0.4 TWh/yr assuming a cost less than 10 p/kWh at an 8% discount rate.

Geothermal

The UK does not possess any high enthalpy geothermal resources. Research into
hot dry rocks at Rosemanowes Quarry, Cornwall, started in 1976 but was stopped

Table 4
Trends in renewable energy supply and use

	Unit	1990	1992	1995	1996	2000	1990-1996 (%)[1]	1996-2000 (%)[1]
Renewable TPES (excl. hydro)	**ktoe**	**634**	**756**	**1173**	**1307**	**940**	**12.8%**	**-7.9%**
Percentage of TPES	%	0.3	0.3	0.5	0.6	0.4		
Geothermal	ktoe	0	0	0	0	0	n.a.	n.a.
Solar, Wind, Wave, Tide	ktoe	1	3	34	42	40	94.4%	-1.1%
Biomass and Wastes[2]	ktoe	633	754	1140	1265	900	12.2%	-8.2%
– Biomass	ktoe	468	561	674	723	n.a.	7.5%	n.a.
– Wastes	ktoe	165	193	466	542	n.a.	22.0%	n.a.
Renewable electricity generation (excl. hydro)	**GWh**	**1393**	**2154**	**5648**	**6007**	**3800**	**27.6%**	**-10.8%**
Percentage of total generation	%	0.4	0.7	1.7	1.7	1.0		
Geothermal	GWh	0	0	0	0	0	n.a.	n.a.
Solar, Wind, Wave, Tide	GWh	9	34	391	486	500	94.4%	0.7%
Biomass and Wastes[2]	GWh	1384	2120	5257	5521	3300	25.9%	-12.1%
– Biomass	GWh	n.a.	n.a.	n.a.	n.a.	n.a.	n.a.	n.a.
– Wastes	GWh	n.a.	n.a.	n.a.	n.a.	n.a.	n.a.	n.a.
Renewable TFC (excl. hydro)	**ktoe**	**215**	**234**	**420**	**427**	**100**	**12.1%**	**-30.4%**
Percentage of TFC	%	0.1	0.2	0.3	0.3	0.1		
Geothermal	ktoe	0	0	0	0	0	n.a.	n.a.
Solar, Wind, Wave, Tide	ktoe	0	0	0	0	0	n.a.	n.a.
Biomass and Wastes[2]	ktoe	215	234	420	427	100	12.1%	-30.4%
– Biomass	ktoe	201	222	337	343	n.a.	9.4%	n.a.
– Wastes	ktoe	14	13	84	83	n.a.	34.3%	n.a.
Hydro TPES	ktoe	445	463	416	289	400	-6.9%	8.5%
Hydro electricity generation	GWh	5171	5383	4838	3361	4650	-6.9%	8.5%
Percent of total generation	%	1.6	1.7	1.5	1.0	1.3		

Notes:

1. Annual Growth Rate

2. Including Animal Products and Gases from Biomass

in 1993 after the government transferred its efforts to a collaborative (and cheaper) project with France and Germany. A small development programme into geothermal heat has resulted in the use of this source in a district heating scheme in one town (Southampton) in the south of England. No other geothermal heat is utilised, and further development of the limited resource is unlikely.

Passive solar

Unplanned use of incident radiation in buildings is estimated by the government to be equivalent to 150 TWh/year.

UNITED STATES

OVERVIEW OF RENEWABLE ENERGY POLICY

The US is the largest energy user among IEA countries, and is also the largest user of non-hydro renewable energy. Total non-hydro renewable energy use was 96.4 Mtoe in 1996, (4.5% of total energy supply), up from 94.5 Mtoe in 1995. The Department of Energy (DOE) aims to support policies capable of doubling non-hydroelectric renewable generating capacity by 2010 as part of its strategic plan to provide energy security, national security, environmental quality, and science leadership[63]. In addition, many states have initiated programmes to promote renewable energy.

At a national level, renewable energy is promoted via a number of different policy measures, such as economic and fiscal incentives (e.g. those included in the 1992 *Energy Policy Act*), and voluntary measures (e.g. those outlined in the 1993 *Climate Change Action Plan*). More recent measures include federal guidelines for electricity restructuring which include suggestions for a renewable portfolio standard, which would mandate a portion of total electricity generation met by renewables.

Non-hydro renewable electricity generation represented 2.3% of total electricity generation in 1996, and hydropower contributed a further 9.6%. Government and legislative support, e.g. via investment or tax credits, is expected to lead in the long-term to a significant increase in capacity of non-hydro renewable energy. Far reaching changes in the electricity sector that are expected with electric utility deregulation are likely to have significant effects on the development and commercialisation of renewables in the US. Federal and state restructuring decisions will be critical to achieving US renewable goals, but in the near term low electricity prices may limit the growth of renewable electricity generation.

POLICIES

US energy policy has changed substantially in recent years with the emergence of issues such as global climate change, clean air, and electric utility industry restructuring. US energy policy goals include protection from oil disruption, investment in R&D to raise long-term productivity and high-wage jobs, a healthier environment, and a balanced portfolio of energy options. The national energy

---

63 US Energy Policy is described and analysed in detail in the IEA's *Energy Policies of The United States, 1998 Review.*

strategy supporting these goals has evolved against a background of changing energy prices; shifting political commitment to renewable energy promotion; emerging environmental imperatives; and the emergence of broader forces reshaping the electric utility sector such as planning, competition, and restructuring.

Federal energy programmes are run by the DOE, and include those designed to improve the cost-effectiveness and efficiency of renewable energy technologies and to reduce the cost of renewable energy production. The specific technology and cost-of-energy goals of the programmes and the projected installed capacity for 2000 are summarised in Table 1.

Restructuring the electric utility industry to allow a more open, competitive market is either underway or is planned in all but three States by January 1998. To date, the federal legislation proposed to guide restructuring focuses mostly on better defining state and federal powers, reducing federal interference in state efforts by repealing laws such as PURPA, and providing protection for consumers. Restructuring can also be used to create a supportive market mechanism, a

Table 1
1998 Programme for Renewable Energy Generation of Electricity

Technology	'Cost-of Energy' Goals	1996 Reported Capacity	Domestic US Projections for 2000 (1)
Photovoltaics	12¢-20¢/kWh (3)	10 MW	20 MW
Solar Thermal Electric	distributed power: 12¢/kWh dispatchable power: 9¢/kWh (2)	360 MW	360 MW
Wind	2.5¢/kWh by 2002 (2)	1,850 MW	2,550 MW
Geothermal	3¢-7¢/kWh (2)	3,020 MW	3,020 MW
Biomass (Gasification Technologies)*	No cost goal; demonstrate feasibility of biomass systems (2)	7,320 MW	7,840 MW
Hydro**	No cost goal; improvement in environmental effects to reverse decline in total US generation (2)	78,580 MW	80,320 MW

Notes: Does not include effects of proposed Renewable Portfolio Standards.

*   Includes co-generation.

**   Does not include pumped storage.

Source:

(1) *Annual Energy Outlook 1998*, Energy Information Administration, US Department of Energy.

(2) Fiscal Year 1999 Program Briefs, Office of Utility Technologies, US Department of Energy.

(3) Photovoltaics the Power of Choice, National Photovoltaics Program Plan for 1996-2000, US Department of Energy.

renewable portfolio standard (RPS), which would mandate that a portion of all electricity is produced from renewables. Guidelines released by the administration include an RPS of 5.5% for non-hydro renewables. As of January 1998, of the 16 state-led pilot projects to test electric competition across the nation, ten states had enacted restructuring legislation and six had issued comprehensive regulatory orders. Eight of the ten states that have enacted laws also include an RPS or Systems Benefits Charge to support renewables.

COP-3, which was held in December 1997 in Kyoto, Japan, is also likely to affect the future development of renewable energy in the United States. The Kyoto Protocol calls for the United States to reduce emissions to 7 percent below 1990 levels by the period 2008-2012, although not all reductions are required to come from the energy sector. However, the US government has not yet ratified this Protocol.

The DOE is responsible for overseeing federal research on renewable energy technologies. The implementation of these programmes relies on collaboration with key stakeholders. The National Renewable Energy Laboratory (NREL) is responsible for managing the majority of DOE renewable R&D projects, and in turn subcontracts to industry to support a variety of development activities. The Fiscal Year (FY) funding for 1998 for all of DOE's R&D programmes aiming to provide clean energy technology options for electricity supply was 250 M USD. R&D support for solar programs (photovoltaics and solar thermal), at 96 M USD, was the largest portion of this funding, followed by wind, at 32.5 M USD, geothermal (29 M USD), biomass (28.2 M USD) and hydropower (0.75 M USD). An increase of approximately 100 M USD for 1999 (spread over PV, wind, geothermal and biomass) has been proposed by the Clinton Administration.

In addition to DOE's annual programme implementation plans, legislation and other interagency activities also support the development and commercialisation of renewables. There are incentives targeted at increasing the known renewable resource base and for increased renewable energy production. A summary of such laws and plans is provided below.

The *Public Utility Regulatory Policies Act* (PURPA), passed in 1978, amended the Federal Power Act by requiring utilities to purchase power from certain 'qualifying' non-utility producers, especially small (below 80 MW) renewables-based electricity production, at 'avoided cost' rates. The qualifying facilities were also exempt from some of the state and federal regulations that applied to utility generators. PURPA encouragement of non-utility generation has contributed to increased electricity production from geothermal, biomass, waste, solar, and wind. However, PURPA creates an environment which does not favour renewables as the current low cost of conventional fuels results in an 'avoided cost' which is lower than that of renewables. Restructuring of the US electric utility industry necessitates PURPA reform or repeal to support the new, emerging market structure.

The *Energy Policy Act* of 1992 (EPACT) reformulated the tax incentives available for renewables energy. EPACT supports renewables in three ways:

■ the permanent extension of the investment credit (Section 1916), provides a 10 percent investment credit for most solar technologies and geothermal;

■ the production tax credit (Section 1914) supports wind and closed-loop biomass and is available to investor owned utilities and non-utility generators for up to ten years for electricity produced in stations brought on line before July 1, 1999. The Administration has proposed to extend this credit by 5 years;

■ the production incentive payment (Section 1212, Renewable Energy Production Incentive) is only available to facilities which cannot avail themselves of the tax credit because they pay no federal income taxes (including any facilities owned by a state or any political subdivision of a state or any corporation or association wholly owned by these or by a non-profit electrical cooperative). This payment is available for solar, wind, biomass (excluding MSW) and geothermal (excluding dry steam). Some 5.5 M USD had been paid by FY 1997, under this programe.

The 1993 *Climate Change Action Plan* (CCAP) outlined US plans to stabilise emissions of greenhouse gases to 1990 levels by 2000. A renewables component of the plan aimed to lower the cost of renewable energy development through facilitation of aggregated buyer purchasing, and increase the penetration of renewable energy through application of integrated resource planning. The actions outlined in the CCAP were coordinated by the Department of Energy, the Environmental Protection Agency, the Department of Transportation, US Department of Agriculture, and the Department of State. The *Climate Challenge Programme* (CCP) is part of the CCAP and is a voluntary programme between utilities and DOE. The programme pursues cost-effective actions to reduce, avoid, or sequester greenhouse gases. By January 1997, emissions reductions pledged through the CCP due to increased use of renewable sources amounted to 3.2 Mt carbon (11.7 Mt CO_2).

Economic incentives for renewable-based transport fuels include a federal tax exemption of 3¢/US gallon for fuels containing at least 12.7 percent Ethyl Tertiary Butyl Ether (ETBE). The current federal excise tax structure also allows gasoline blends containing 5.7, 7.7, and 10 percent alcohol to enjoy an exemption of 3¢/US gallon. Policy makers have concluded that the ethanol content of ETBE used in reformulated gasoline should enjoy the same tax exemption, however, there is insufficient ethanol in ETBE to warrant the higher exemption rates for 7.7 and 10 percent blends. The ETBE exemption is the subject of a national policy debate and it is unclear how or where the exemption should be implemented. The DOE's Biofuels Programme aims to develop the technologies that can be major contributors to producing sufficient domestic biofuels to substitute, on an energy equivalent basis, 10 percent by 2000 and 30 percent by 2010 of the projected consumption of motor fuel by light duty vehicles in the US. This penetration of biofuels into Light Duty Vehicles and other vehicles would then result in CO_2

emissions reductions of 146 Mt by 2000 and 292 Mt by 2010 as well as reduced expenditure on oil imports.

In 1994, the US Supreme Court ruled that the *Clean Water Act* of 1977 allows states to impose conditions on hydropower facilities, such as minimum flow requirements. This is likely to result in the decline of the relative importance of hydropower, as the required FERC licensing and relicensing conditions on hydropower operators, where water has to be released over spillways rather than through turbines, may limit exploitation of existing and planned plants and even force closures as with the Edwards Dam in Maine. The DOE hydropower programme is scheduled to conclude a state-by-state hydropower resource assessment in FY 1998.

The states have various financial and regulatory incentive programs to encourage different types of renewable energy use, production, or sales. For example, Virginia has a manufacturing incentive grant of seventy-five cents per watt, up to a total of six million watts. Massachusetts has a local property tax exemption for wind, solar, and hydro units. Massachusetts and Indiana have laws that allow for agreements to protect access to solar resources, but do not make solar access an automatic right. In addition, "green power" is available in many parts of the country.

The Department of Defense is also aiming to increase its own use of renewable energy technologies whenever cost-effective. Since 1995, more than 300 kW of photovoltaics, over 4,000 geothermal heat pumps, and a 7-kW demonstration dish-stirling engine have been installed on US military installations.

The Utility Photovoltaic Group (UPVG) is a not-for-profit association of 91 electric utilities that addresses technical, economic, and policy issues with photovoltaics. UPVG manages the TEAM-UP programme (Technology Experience to Accelerate Markets for Utility Photovoltaics), which provides matching federal funding for both grid-connected and grid-independent utility projects to develop sustainable relationships between the photovoltaic manufacturing and electric utility industries.

California has implemented the largest restructuring programme in the United States. Its programme includes a non-bypassable systems charge to pay for transmission maintenance, public purpose programs, and for the support of renewable energy research, development, and deployment. Estimates are that the fund will reach between 465 and 540 M USD for renewables in the first year. The first four years of the programme will result in support for renewables averaging 45 percent for existing technologies (through tax credits and capacity payments), 30 percent for new technologies (through competitive bids), 10 percent for emerging technologies (through multiple competitive RFPs), and 15 percent for the customer (through a proposed per kWh consumption credit to support marketers, aggregators, or suppliers who sell directly to end-use consumers).

STATUS OF RESOURCE EXPLOITATION

Non-hydro renewables contributed 96.4 Mtoe (4.5%) to total primary energy supply in the US in 1996. While the majority (54.5 Mtoe) was used in energy and transformation industries (for the generation of electricity and heat), 41.8 Mtoe was used directly. Biomass and wastes together account for over 70% of non-hydro renewable energy supply, and geothermal the majority of the remainder. Around 2.3% of electricity was generated from non-hydro renewable energy sources in 1996, with a further 9.6% from hydropower.

The proportion of non-hydro renewable electricity has dropped slightly over recent years, reflecting small drops in wind and geothermal generating capacity. However, this trend is expected to be reversed and capacity to grow significantly between 1996 and 2010 (see Tables 2 and 3). Exploitation of non-hydro renewables is concentrated mainly in the Northwest, West and South Atlantic, where over three-quarters of the total renewable capacity is installed. California accounts for 90% of the US's geothermal capacity and 95% of the solar capacity, and has over 16,000 wind turbines in operation.

Table 2
**Electricity Capacity and Generation of Renewable Energy Sources,
1996 and 2010**

Technology	1996		2010	
	Capacity (GW)	Generation (TWh)	Capacity (GW)	Generation (TWh)
Biomass/Wood Waste*	7.3	44.2	8.2	50.8
Geothermal	3.0	15.7	2.9	17.6
Solar**	0.4	0.9	0.7	1.8
Hydropower***	76.5	350.9	80.7	318.7
Wind	1.7	3.4	3.3	7.8
Solid Waste*	3.3	19.0	4.4	28.6
Total excl. hydro	**15.7**	**83.2**	**21.4**	**106.6**
Total	**92.2**	**434.1**	**102.1**	**425.3**

Notes: Does not include effects of proposed Renewable Portfolio Standards.

* Biomass and Wood Waste includes co-generation.

** Solar includes solar thermal and solar photovoltaic generation.

*** Hydropower does not include pumped storage facilities.

Source: *Annual Energy Outlook 1998,* Energy Information Administration, US Department of Energy, and IEA databases.

Biomass

Biomass is used in industry as a fuel for electricity generation and heat production. Electricity production from biomass accounted for the majority of the biomass use in the US and has grown slightly since 1992 to stand at 44.2 TWh in 1996. The development of biomass-based electricity capacity has flourished since the introduction of PURPA in 1978. Total capacity of biomass and waste plants pre-PURPA stood at 200 MW[64] but had reached 7.3 GW by 1996. Direct use of biomass accounted for 29.6 Mtoe or 2 percent of 1996 US total energy consumption. Biomass is also used for transportation biofuels, but to a much lesser extent.

Wastes

Electricity generation from municipal solid wastes was 19.0 TWh in 1996. The Energy Information Administration of the DOE projects that electricity generation from municipal solid waste (MSW) will grow faster to 2010 than electricity generation from any other renewable source, with capacity projected to expand to 4.38 GW. However, this increase may be dampened by US Supreme Court rulings that ash generated from an MSW plant must be treated as hazardous waste. In addition to environmental concerns and policies that favour recycling as opposed to incineration, federal tax policies are discouraging the construction of capital intensive facilities by municipalities. The waste-to-energy industry is also feeling the impact of competitive deregulation which is driving down the price of conventionally generated electricity.

Wind

Approximately 1.7 GW of wind capacity was installed at the end of 1996. This capacity is lower than in previous years as retirements have outpaced additions. Estimated generation has increased by 12 percent from 3.0 TWh in 1993 to 3.4 TWh in 1996, and is projected to increase substantially by 2010. Current costs are generally quoted at 4 to 6 cents/kWh, depending upon financing and wind resource. The US Department of Energy's Wind Energy Programme has a 2.5 cents/kWh cost-of-electricity goal by 2002 for good wind sites with favorable financing.

Solar

Electricity generation from both solar thermal and solar photovoltaic power reached an estimated 0.906 TWh in 1996 (compared to less than 0.02 TWh in

---

64 D.L. Klass *Biomass Energy in North American Policies*, in *Energy Policy*, Volume 23, No. 12, December 1995.

1986). This output is largely from the 0.35 GW of solar thermal troughs operating in California. Growth in the early part of the 1990s has been slow, but a number of events are expected to increase both solar thermal and solar photovoltaic generation in the United States. Solar Two, a 10 MW power tower, was commissioned in 1996 and is expected to increase its generation as improvements in design and storage are realised. The TEAM-UP program, previously described, began awarding contracts in 1995 and is in an active deployment phase. Over 3 MW have since been deployed and future commitments of 5 MW will add a total of 8 MW by the year 2000. Prior to COP-3, the Administration announced the Million Solar Roofs Initiative, which has a goal of installing over one million residential and commercial solar installations. Of these, sixty percent (an estimated 1.8 MW) would be solar photovoltaic and the remaining forty percent solar hot water systems.

The deployment of photovoltaics has grown significantly as a result of the combination of co-operative federal/industry programs, favorable state-level policies, and federal financial support. The largest federally-operated grid-connected photovoltaic station, PVUSA in Davis, California, was initiated in 1987 to demonstrate various photovoltaic technologies for bulk generation by electric utilities. The federal government continues to provide a significant portion of funding for PVUSA. However, management of the facility was transferred in 1997 to the California Energy Commission and the Sacramento Municipal Utility District.

Direct use of solar energy was estimated at 1 Mtoe in 1996.

Hydro

Electricity generation from hydropower provided 9.6% of total electricity in 1996 (350.9 TWh), compared to between 250 and 300 TWh/year in most years since 1970. Capacity is not expected to increase substantially in the long-term, and growth in hydro power is not expected to keep pace with the modest anticipated growth in overall electricity demand. The US Department of Energy is focusing its hydropower R&D efforts on developing advanced turbine technology which will minimise the environmental effects of both existing and new projects. This work will facilitate both re-licensing and re-powering of plants with environmental concerns.

Geothermal

Geothermal generation in 1996 was 15.7 TWh, reversing a declining trend that was first recognised in the late 1980s. Geothermal generation is expected to continue increasing to 2010. At present, only hydrothermal resources are exploited commercially. However, the basic technology for extracting hot dry rock energy has been proven at a test site in New Mexico and efforts are underway to increase commercialisation of enhanced geothermal systems (both dry and wet systems on the margins of known and/or exploited resources). In addition to electricity generation, geothermal energy was used for direct use applications. Use in 1996

Table 3
Trends in Renewable Energy Supply and Use[1]

	Unit	1990	1992	1995	1996	2000	1990-1996 (%)[2]	1996-2000 (%)[2]
Renewable TPES (excl. hydro)	**ktoe**	**76280**	**88467**	**94525**	**96365**	**103364**	**4.0%**	**1.8%**
Percentage of TPES	%	4.0	4.5	4.5	4.5	4.5		
Geothermal	ktoe	13770	14764	22599	24292	27719	9.9%	3.4%
Solar, Wind, Wave, Tide	ktoe	255	315	1346	1371	2327	32.4%	14.1%
Biomass and Wastes[3]	ktoe	62255	73387	70579	70703	73318	2.1%	0.9%
– Biomass	ktoe	n.a.	63784	61059	60915	n.a.	n.a.	n.a.
– Wastes	ktoe	n.a.	9603	9520	9788	n.a.	n.a.	n.a.
Renewable electricity generation (excl. hydro)	**GWh**	**84842**	**86321**	**81882**	**83299**	**108999**	**-0.3%**	**7.0%**
Percentage of total generation	%	2.7	2.6	2.3	2.3	2.8		
Geothermal	GWh	16012	17168	14941	15746	18278	-0.3%	3.8%
Solar, Wind, Wave, Tide	GWh	2962	3666	4024	4316	6704	6.5%	11.6%
Biomass and Wastes[3]	GWh	65868	65487	62917	63237	84017	-0.7%	7.4%
– Biomass	GWh	n.a.	43591	43876	44203	56566	n.a.	6.4%
– Wastes	GWh	n.a.	21896	19041	19034	27451	n.a.	9.6%
Renewable TFC (excl. hydro)	**Ktoe**	**22592**	**31527**	**39518**	**41838**	**43916**	**10.8%**	**1.2%**
Percentage of TFC	%	1.7	2.4	2.8	2.6	2.9		
Geothermal[4]	ktoe	0	0	9750	10750	12000	n.a.	2.8%
Solar, Wind, Wave, Tide[4]	ktoe	0	0	1000	1000	1750	n.a.	15%
Biomass and Wastes[3]	ktoe	22592	31527	28768	30088	30166	4.9%	0.1%
– Biomass	ktoe	n.a.	31527	28690	29663	n.a.	n.a.	n.a.
– Wastes	ktoe	n.a.	0	77	424	n.a.	n.a.	n.a.
Hydro TPES	ktoe	23491	21876	27014	30180	27447	4.3%	2.3%
Hydro electricity generation	GWh	273152	254371	314116	350932	319156	4.3%	2.3%
Percent of total generation	%	8.6	7.8	8.8	9.6	8.1		

Notes:

1. Renewables data formally submitted on an annual basis by country's administration to the IEA have in some instances been supplemented with national publications or other sources.

2. Annual Growth Rate

3. Including Animal Products and Gases from Biomass

4. Data unavailability before 1993 for direct use of geothermal and solar energy means that there is a break in series in the above data. Calculated growth rates for TPES and TFC therefore overstate actual growth rates.

Source: *Annual Energy Outlook* 1996, 1997, and 1998, Energy Information Administration, U.S. Department of Energy.

was 1.67 TWh (thermal) and is projected to increase to 4.2 TWh by 2010. Space heating is the largest direct use application.

Heat pumps

Geothermal heat pumps (GHPs) provide space heating, cooling, and water heating for commercial, industrial and residential buildings. Because GHPs draw thermal energy from the earth rather than the atmosphere, they are much more efficient than ordinary air source heat pumps. Although the initial cost for GHPs is higher, their operating costs are markedly lower than conventional heating, making GHP economics attractive on a life cycle basis. The electricity savings gained by the use of geothermal heat pumps for both residential and commercial building heating grew by 52 percent from 2.3 TWh in 1994 to 3.5 TWh in 1996. These savings are projected to increase to 38.4 TWh in 2010.

EUROPEAN UNION

OVERVIEW

The 1995 White Paper on energy policy identified protection of the environment as one of the three key energy policy issues facing the European Union. This was followed by a Green and then White Paper on renewable energy development in the EU. This renewable energy White Paper includes a target of 12% for renewable energy's importance in the Community's energy mix by 2005, compared to its current level of under 6%.

Support for renewable energy development at an EU level (i.e. in addition to programmes undertaken at a national level) has been available for many years through different EU programmes. Environmental commitments, particularly the proposal to cut emissions of greenhouse gases, and the long-term outlook for energy supply have led to a recent strengthening of these programmes, with the development of ALTENER, a programme devoted exclusively to promoting the non-technical barriers to the use of renewable energy.

Other Community actions not specifically directed towards renewables could, nevertheless, have a significant impact on their future use. In particular, the creation of the single energy market could have signifiant impacts on fuel inputs to electricity generation.

Much of the Community action on renewables is directed at improving the flow of information: this is addressed specifically in the ALTENER programme. In addition, the results of research undertaken under other programmes, such as JOULE-THERMIE, as well as information on research, programmes and publications financed or co-ordinated by the EU are widely disseminated, including on the Internet.

POLICIES

The Energy Policy objectives of the European Union are stated in the December 1995 White Paper *An Energy Policy for the European Union* and are based on the three key energy policy objectives of competitiveness, security of supply and environmental protection. The EC's November 1996 Green Paper on Renewable Energy outlined other policy goals to which increased use of renewable energy sources can help. These include regional development, increased employment, social and economic cohesion, increased competitiveness, as well as protection of the environment and security of energy supply through the diversification of our energy sources.

The EU pursues three broad types of measure to promote renewable energy: co-ordination and dissemination of information; financial incentives (including for information dissemination, development of standards and other) and R&D. It has also set a target for renewable energy use.

The Commission's work on renewable energy is concentrated in DGXVII (Energy) and DGXII (Science, Research and Development), although the cross-cutting nature of renewable energy means that DGXI (Environment), DGVI (Agriculture) and others are also involved in some of the programmes.

There are several on-going Community programmes in the field of energy[65], of which one (ALTENER, and its sucessor, ALTENER II) is devoted solely to renewable energy. Other energy programmes such as JOULE/THERMIE and SYNERGY have a renewable energy component (see Table 1). Renewable energy projects can also benefit from Community programmes or funds that are not specifically directed at the energy sector, such as those of the European Investment Bank or of the MEDA programme. Some of these programmes are also open to non-Community Member States. The remainder of this chapter will deal with energy-related programmes only.

ALTENER I and II aim to increase the use of renewable energies by overcoming the non-technical barriers to their use, e.g. by creating the necessary legal, socio-economic and administrative conditions for the implementation of an action plan for renewables, and by encouraging private and public investments in the production and use of energy from renewable sources. The objectives of the ALTENER programmes are to contribute to achieving the overall Community objectives and concerns of limiting CO_2 emissions, increasing the share of renewable energy sources in the energy balance, reducing energy import dependence, ensuring the security of energy supply, and fostering economic development, economic and social cohesion and regional and local development. The focus of the ALTENER programmes are therefore the countries that make up the EU, although the programmes are open to co-operation from the Central and Eastern European countries who have asked to become members of the EU.

Renewable energy sources included in the ALTENER programmes include biomass, thermal and photovoltaic solar systems, solar systems in buildings, small-scale (<10 MW) hydroelectric projects, wind power and geothermal energy. ALTENER I (1993-1997) had a budget of 40 MECU. The specific objective of the ALTENER I programme was to achieve a 180 Mt reduction in CO_2 emissions by increased renewable energy use. ALTENER II (1998-2002) has a proposed budget of 22 MECU for the first two years. Actions and measures relating to renewable energy sources which will be financed under ALTENER II include: targeted projects

---

65   More information is available on the EU web server: http:/europa.eu.int/eu/comm/dg17/dg17home.htm.

to facilitate and prepare investments, information and training (including in local/regional planning); development of harmonised standards for renewable energy technologies; development of new financial arrangements, and improved co-ordination between renewable energy initiatives in different parts of the Community.

The JOULE/THERMIE programme (1995-1998) is a technical programme concerned with non-nuclear energy (renewable energy, rational use of energy and fossil fuels), and follows on from the (separate) JOULE and THERMIE (1990-1994) programmes. The objectives of this programme include promoting the efficient use of indigenous resources of energy and reducing environmental emissions, in particular CO_2. The JOULE part of the new programme focuses on R&D, and the THERMIE part of the programme focuses on demonstration and deployment. Community support covers a maximum of 40% of the total eligible costs of demonstration projects or 50-100% of costs for other projects (such as pre-feasibility studies). The projects have to be proposed by a consortium of bodies from two or more Member states. The overall aim of the programme is to stimulate the introduction and integration of renewable energy sources that offer substantial advantages from the point of view of environmental protection, CO_2 emissions, and security of energy supply. The budget for the four-year programme is 1030 M ECU, of which an indicative 45% should be allocated to renewable energy sources.

The basic objective of SYNERGY is to improve long-term Union energy security by helping other countries to make effective energy policy decisions. The environmental impact of energy use is also an important consideration. The projects financed therefore focus on actions that foster policy dialogue, help policy development and develop energy institutions, e.g. conferences and Energy Centres. The actions undertaken in SYNERGY focus on the energy sector and therefore include a renewables component.

The current target for renewable energy is that it should account for 12% of the Community's energy supply by the year 2010. This is an increase of an earlier target, outlined in ALTENER I, of 8% of total energy supply by 2005. The reasons behind the increased target are not only the recent success of renewable energy programmes in some EU member states, but the increased emphasis on cleaner energy supply, as outlined in the 1995 *White Paper* on energy policy. This emphasis is also translated into the total budget for ALTENER II, which is likely to be higher than that for ALTENER I.

Other Community actions, such as opening up the internal market for electricity, could have important ramifications for fuel diversity. The internal market should ensure that electricity producers can choose the most appropriate and cost-effective fuel for electricity generation.

In the fiscal field, Member States can already apply exemptions or reduced rates of excise duty in the field of pilot projects for the technological development of more environmentally-friendly products and may, moreover, seek authorisation from the

Council to introduce further exemptions or reductions for specific policy considerations (Article 8(2)(d) of Council Directive 92/81/EEC). The Commission will also issue a Communication which will analyse the contribution that the use of fiscal instruments can make to the promotion of energy efficiency and renewables, taking into account general policy in the fiscal field.

In addition, a strategy to promote Third Party Financing (TPF) will be developed in order to improve energy efficiency, particularly at regional and local level and to encourage the market penetration of innovative technologies, products and services.

In this area, the local level, notably regions and cities, have an important role to play in exchanging experience, know-how and technology transfer, since for many renewable technologies, it is necessary to have an in-depth knowledge of local situations.

Table 1
EU and other Programmes concerning Renewable Energy

PROGRAMME	BRIEF DESCRIPTION	BUDGET	CONTACT
EU FOURTH FRAMEWORK PROGRAMME FOR RTD (4FP)	19 R&D programmes among which non-nuclear energy.	13000 MECU	Mr A Damiani, DGXII
JOULE (-THERMIE)	Non-nuclear energy R&D	1030 MECU	Mr M Poireau DGXII
(JOULE-)THERMIE	Non nuclear energy – Demonstration projects – Dissemination		DGXVII Mr E. Millich Mrs. P.-L. Koskimäki
International Co-operation (2nd Activity of 4FP)	Co-operative RTD activities with 3rd countries and International Organisations		For energy topics: R&D: P.Valette (DGXII) Demonstration (THERMIE) Ms P. Marques (DGXVII)
Innovation Programme (3rd activity of 4 FP)	Dissemination and exploitation of RTD results (includes "OPET" network)		Mr J. Hernandez-Ros DGXIII
ALTENER - Actions for greater penetration of Renewable Energies in Europe	Studies, evaluations and information ("Article 3" type activities)	45 MECU	Mrs. M-A. Perez Latorre EC/DGXVII-C-2
SAVE	Increasing Energy Efficiency, basically a framework for EC legislation and standards for energy efficiency	35 MECU	Mr. E. Dalamangal, DG XVII

Further to the Commission's aim to develop an integrated R&D strategy, and its support to renewable R&D via the different programmes described above, it has compiled a database and atlas of current R&D activities that are being undertaken at the European level (CORDIS)[66], including renewable energy-related R&D (Table 2).

Table 2

Databases Relevant to THERMIE/Renewable Energy Technology

DATABASE NAME	BRIEF DESCRIPTION	RESPONSIBLE/CONTACT
SESAME	Energy Demonstration Project Records	Info Partners s.a. Fax. +352.3498.1234
ISET (Information System on Energy Technology)	CD-ROM for Microsoft Windows (3.1/3.11)	ICEU, Leipzig Fax. +49.341.980.3486
CEFENE (Energy Efficient Equipment manufactured in EU)	Database (based on 'Kompass' Directory) on CD-ROM	EUROPLAN (F) Fax. +33.93.74.31.31
Member States' and International Organisations' Programmes	Internal OPET-CS Database – In preparation	OPET coordination unit opet_cu@ecotec.com
World Wide Web – CORDIS – THERMIE on the Internet	– EU RTD Information System **http://www.cordis.lu/** – Information brochure on THERMIE	CORDIS Help Desk E-mail helpdesk@cordis.lu –

---

66   Available at http:/www.cordis.lu.

11

ANNEXES

ANNEX A

HYDROPOWER

Introduction

Hydropower is a form of renewable energy that has been harnessed for more than 100 years in some IEA countries. Hydropower supplied 111 Mtoe (2.4%) of the IEA's total energy supply in 1996, and 15.9% of the IEA's total electricity requirements in the same year, and is the second most important renewable energy at the IEA level after biomass. Hydropower is produced when the kinetic energy of water flowing through a turbine is transformed to electric energy. The amount of electricity produced from a turbine is proportional to the volume of water flowing through it, and on the height difference ("head") between the reservoir intake and the water outlet in the power plant.

Electricity can be generated from "run of river" systems (where there is no reservoir). However, the majority of the IEA's hydropower is generated from systems with a lake/reservoir: a water reservoir enables the water in that reservoir to be used for power generation on demand, rather than only when river flow is over a certain level. This temporal storage of water (and therefore of electricity generating capability) is important on a daily, and even seasonal, basis, as electricity demand may not be in step with precipitation/river flow patterns. For example, reservoirs are often at their fullest in spring, when rainfall is augmented by snow thaw, although electricity demand may be highest in winter, when it is coldest.

The timing and quantity of electricity produced from a hydropower plant can be varied by the operator to cover electricity consumption in peak periods (within the limits of water availability and inflow). However, a plant cannot generate electricity at maximum capacity continuously throughout the year because the water used to generate electricity has to be replaced (either naturally via precipitation or snow thaw, or by pumped storage). Production is therefore intermittent, although predictable over short time horizons. This is unlike electricity production from wind and solar electricity, where minute-by-minute variations can be significant, and where it is difficult to forecast precise electricity production patterns more than a few hours ahead. It is also different from production patterns in thermal stations, which can operate continuously, given sufficient quantities of fuel.

Status and Prospects

Notwithstanding climatic variations, total hydropower production in the IEA has increased steadily since the 1970s. However, the proportion of the IEA's electricity consumption met by hydropower has dropped from over 23% in 1971 to under 16% in 1996. This is largely because hydro capacity (excluding pumped storage capacity[67]) has grown more slowly than total electricity generating capacity. This is, in turn, a reflection of the maturity of hydropower development: the best sites have already been exploited, and expansion potential is limited. Therefore, although generation per unit installed hydro capacity has increased, this increased productivity from existing hydropower units has been at a slower pace than growth in electricity production from other sources.

Hydropower is the major source of electricity production in Norway, New Zealand, Austria, Canada and Switzerland. It also supplies over 30% of total electricity in Portugal, Sweden, and Turkey, and over 15% in four other IEA countries (see figure A1). However, variations in precipitation patterns means that an individual country's hydropower production can vary significantly between years. For example, 1996 was a dry year in northern Europe, and Sweden's hydropower generation dropped to 51.4 TWh compared to 72.5 TWh in 1990 (from approximately the same capacity).

Electricity produced from hydro is still growing in absolute amounts. However, the relative importance of hydropower is expected to decline to under 12% of the IEA's total electricity production by 2020 (see figure A2) as growth in hydropower capacity is outpaced by growth in other electricity generating sources. This is for a number of reasons: firstly, the most economic large hydro sites have already been exploited. Secondly, the remaining economically-viable large hydro sites may not be exploited for environmental reasons: development of a large hydro system and the consequent flooding of land, and changed water flow patterns has significant impact on the local ecosystem. Large hydro developments may also be opposed on other grounds, such as for aesthetic reasons. Thirdly, market reform within the electricity sector, combined with consumers' increasing ability to choose their electricity supplier, means that the financial risks of building large, capital-intensive plants are higher than before as there may be no guaranteed market for the electricity generated.

The majority of hydropower generation in the IEA is produced from large hydro systems. However, since the definition of "large" varies from country to country and since the statistics collected by the IEA do not distinguish large and small

---

67 "Pumped storage" is when off-peak electricity is used to pump water up to reservoirs and this stored water is used to satisfy peak electricity demand. Pumped storage hydro is excluded from "renewables" because the electricity used to pump water up to the reservoir is, in many IEA countries, of non-renewable energy origin (e.g. baseload coal-fired plants). If renewable electricity is used to pump water up to the reservoir, however, pumped storage hydro should be considered as a renewable electricity source.

Figure A1
Proportion of hydropower in total electricity generation, 1994-1996 (ave, %)

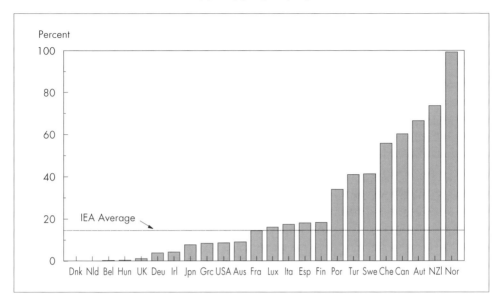

Source: IEA databases.

Figure A2
Trends in hydropower generation and importance, 1973-2020 (TWh and %)

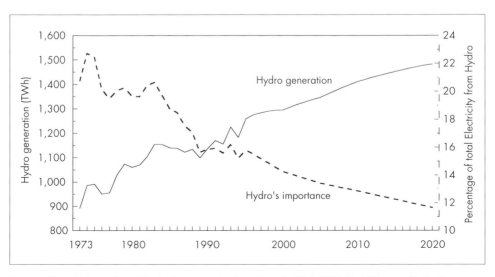

Source: Historical data from IEA databases, projections from the IEA's 1998 *World Energy Outlook*.

hydro, an IEA-wide analysis of small and large hydro trends is not possible. Nevertheless, increased development of small hydro systems are specifically encouraged in many IEA countries, and the relative importance of small hydro plants within the IEA is therefore likely to increase over the forecast period.

Barriers to Increased Hydro Use

The variable cost of hydropower production from existing sites is low. However, the capital-intensive nature and long lead-time for large hydro projects may make cost a significant barrier for new large hydro developments (although the cost of hydro plants are highly site-specific). For example, New Zealand have estimated new "low-cost" hydro will cost 6.7 cents/kWh, compared to 4.3 cents/kWh for new combined cycle gas turbine plants. Sweden also estimate that new hydro development could be more expensive than new fossil fuel plants[68]. Other barriers to hydropower development may be environmental (e.g. relating to water flow or aesthetic disturbance) or administrative (e.g. the time required to obtain the necessary development and operational licences). Nevertheless, some large hydro projects are being planned. For example, in March 1998, Québec and Newfoundland signed an agreement which sets guidelines for subsequent negotiations of a hydro project development at two sites – at Churchill Falls and Gull Island – to provide 3,200 MW of power. Negotiations are not yet finalised.

Although the costs of new large hydro developments may be high, existing plants can be upgraded to increase production capabilities. This is a lower cost option than new development, is subject to fewer objections on environmental grounds, and may also entail a lower administrative burden than new developments. Some IEA countries have been actively pursuing policies to upgrade their existing hydro capacity. For example, Norway introduced subsidies from 1991-1994 to encourage more rapid upgrading of small and medium-sized plants.

Many IEA countries, even those undergoing market reform in their electricity sector, have legislation in place that ensures a guaranteed market for some sources of renewable electricity[69]. However, this legislation, as well as other mechanisms (such as those outlined below) put in place to encourage increased use of renewable energy sources often excludes large hydro developments even if small hydro projects are promoted.

The high level of capital required for a large hydro development, combined with its economic risks, perceived environmental impacts and administrative burdens means that large hydro development is limited in most IEA countries. An exception is Turkey, where electricity demand is growing at 9% p.a. and total hydro capacity is

---

68 More details are given in the country chapters.

69 This is discussed more fully in the overview section, and in the individual country chapters.

projected to grow from 10.1 GW in 1996 to 24.5 GW in 2010: this is more than the projected growth in hydro capacity in Austria, Canada, Japan and Norway combined.

Promotion of Small Hydro in IEA Countries

The current importance of hydropower in some IEA countries' electricity generating mix means that hydropower has already played a significant role in limiting CO_2 emissions from electricity generation (which typically account for a third of energy-related CO_2 emissions in IEA countries), and in increasing countries' energy security and diversity. As with other renewable energy sources, increased use of hydropower can help countries to further improve their energy self-sufficiency and to maintain (or lower) emissions from electricity generation.

One of the reasons that renewable energy is promoted in IEA countries is because its uptake would be much slower without such promotion. Hydropower (particularly in large scale applications) has already achieved a prominent place in many IEA countries' electricity supply mix without needing specific promotional measures. However, there is still a significant potential for small hydro development in IEA countries that has not yet been exploited.

Not all countries choose to encourage all renewable electricity sources, and, as for other renewable electricity sources, hydro plants are not specifically encouraged in all IEA countries. The IEA countries that do encourage increased use of hydropower may do so in a number of ways, including encouraging more efficient use of existing (mainly large hydro) sites. In addition, increased use of small hydro plants is specifically promoted in certain IEA countries, including Japan, Portugal, Greece, Ireland, Finland and the UK because although the absolute costs of small hydro developments are smaller than those for large hydro, the relative costs (per kWh) may be significantly higher.

Policies used to promote small hydro are similar to those used to promote other renewable energies[70], and include R&D (amounting to $1.69m for 9 IEA countries in 1996), economic and fiscal incentives (e.g. Japan), guaranteed markets for small hydro electricity (e.g. Portugal), premium payments for hydropower (e.g. the UK), and regulations (e.g. Greece, which has removed restrictions for the exploitation of small water falls). Alternatively, promotion of small hydro plants may form part of the incentives available for renewable energy promotion in general, e.g. via guaranteed electricity markets. Country-specific policies used to promote hydropower are outlined in the relevant country chapters of this publication.

---

70 Individual policy types are discussed in detail in *Renewable Energy Policy in IEA Countries, Volume I: Overview*, IEA, 1997.

Conclusions

Hydropower represents an indigenous form of energy that can be used in mature technologies to lessen the environmental impact of energy use. Nevertheless, growth in hydro generation and capacity is slower than that for many other renewable electricity sources. This is partly because the technology is mature and that the best sites have already been exploited. In addition, environmental objections or other barriers to large hydro plants means that many of the remaining potential hydro sites in IEA countries will remain undeveloped.

Small hydro systems are actively promoted in many IEA countries, and the importance of small hydro plants has grown and is likely to continue doing so. The policies used to promote small hydro systems are similar those used to promote increased use of renewable electricity from other sources, although some countries explicitly exclude hydropower or large-scale hydropower from their policies used to promote renewable energy. As for other renewable electricity sources, guaranteed markets combined with premium prices have resulted in capacity and generation from small hydro systems growing rapidly in some countries. However, the IEA's aggregation of data from small and large hydro plants does not permit an IEA-wide analysis of these trends.

ANNEX B

RENEWABLE ENERGY POLICIES IN PLACE IN EACH IEA COUNTRY

	Electricity Supply Industry		Green pricing (not nationwide in any country)	Economic or fiscal incentives	Regulatory Measures/ Standards	Information and Education	Other targets or quotas	Voluntary Actions
	Output/ capacity targets or quotas	Favourable/ Guaranteed markets						
Australia	✔		✔	✔ 2,3	✔ 3	✔ 4		
Austria		✔	✔	✔ 1,2		✔ 4		✔
Belgium				✔ 1,2,3,5	✔ 1,2,4	✔ 1,4	✔ 7 (Flan)	
Canada				✔ 1,3		✔ 1,2		✔
Denmark	✔ cap	✔ f/g	✔	✔ 1,5	✔ 1,4 no landfill	✔	✔ 1,2	
Finland				✔ 1,3	✔ 1,2,4	✔ 1,2,4	✔ 5	✔
France	✔ cap.	✔ fav-wind only		✔ 1,3,4 (low VAT)	✔ 4	✔	✔ 3,6 (wood)	
Germany		✔ fav/g	✔	✔ 1,2,3,4	✔ 1,3	✔ 1,3,4		✔
Greece	✔	✔		✔ 1,3		✔ 1,4	✔ 1,5	
Hungary				✔ 1,2				
Ireland	✔ capacity	✔		✔ 1,3		✔ 1,3,4		
Italy	✔ cap.	✔		✔ 3	✔ 3			
Japan	✔ capacity	✔ f		✔ 1,2,3	✔ 1	✔	✔ 1,6	✔
Luxembourg		✔		✔ 1,3		✔		
Netherlands	✔ output	✔ (< 8MW)	✔	✔ 3,5	✔	✔ 1,3,4	✔ 1,7	✔
New Zealand				✔ 3	✔ 4	✔ 1,5		
Norway				✔ 1, (3 planned)		✔ 4		✔

	Electricity Supply Industry		Green pricing (not nationwide in any country)	Economic or fiscal incentives	Regulatory Measures/ Standards	Information and Education	Other targets or quotas	Voluntary Actions
	Output/ capacity targets or quotas	Favourable/ Guaranteed markets						
Portugal	✔ capacity	✔		✔ 1,2,3	✔ 3	✔ 1,4		
Spain*	✔ both	✔		✔ 1			✔	
Sweden		✔ g/f	✔	✔ 1,3,5	✔			
Switzerland	✔	✔	✔	✔ 1		✔	✔ 5,7	✔
Turkey	✔ cap.			✔ 1,2	✔	✔	✔ (regional)	
United Kingdom	✔	✔	✔	✔ 1	✔	✔		✔
United States	✔ (PURPA)		✔	✔ 1,5		✔		✔
EU				✔ 1,(3),4 (R&D)	✔	✔	✔ 7	

* = regular evaluation/progress report of policy

Economic and Fiscal Incentives, e.g:
1. Grants and subsidies involving direct transfers
2. Credit instruments (interest rate loans, soft loans, loan guarantees)
3. Tax exemptions (tax reliefs, credits, deferrals)
4. Others
5. Output credit for renewable electricity (on top of normal electricity buyback rate).

Regulatory Measures and Standards, e.g:
1. Planning/siting legislation
2. Survey requirements or mapping.
3. Building codes
4. Others (generally waste-related)

Information and Education
1. Publications, advertising campaigns
2. Courses for industry
3. Education programmes in schools and workplaces
4. Renewable energy advice centres
5. Others

Other targets
1. Solar heat
2. Passive solar
3. Biofuel
4. Heat pumps
5. Heat production
6. Other
7. Total

ANNEX C

CONVERSION FACTORS, DECIMAL PREFIXES AND COUNTRY ABBREVIATIONS

Table C1
General conversion factors for energy

To:	TJ	Gcal	Mtoe	MBtu	GWh
From:	multiply by:				
TJ	1	238.8	2.388×10^{-5}	947.8	0.2778
Gcal	4.1868×10^{-3}	1	10^{-7}	3.968	1.163×10^{-3}
Mtoe	4.1868×10^{4}	10^{7}	1	3.968×10^{7}	11630
MBtu	1.0551×10^{-3}	0.252	2.52×10^{-8}	1	2.931×10^{-4}
GWh	3.6	860	8.6×10^{-5}	3412	1

Decimal Prefixes

10^{1}	deca (da)	10^{-1}	deci (d)
10^{2}	hecto (h)	10^{-2}	centi (c)
10^{3}	kilo	10^{-3}	milli (m)
10^{6}	mega (M)	10^{-6}	micro (μ)
10^{9}	giga (G)	10^{-9}	nano (n)
10^{12}	tera (T)	10^{-12}	pico (p)
10^{15}	peta (P)	10^{-15}	femto (f)
10^{18}	exa (E)	10^{-18}	atto (a)

Country abbreviations used in this report

Aus	Australia	Jpn	Japan
Aut	Austria	Lux	Luxembourg
Bel	Belgium	Nld	Netherlands
Can	Canada	NZl	New Zealand
Dnk	Denmark	Nor	Norway
Fin	Finland	Por	Portugal
Fra	France	Esp	Spain
Deu	Germany	Swe	Sweden
Grc	Greece	Che	Switzerland
Hun	Hungary	Tur	Turkey
Irl	Ireland	UK	United Kingdom
Ita	Italy	USA	United States of America

MAIN SALES OUTLETS OF OECD PUBLICATIONS
PRINCIPAUX POINTS DE VENTE DES PUBLICATIONS DE L'OCDE

AUSTRALIA – AUSTRALIE
D.A. Information Services
648 Whitehorse Road, P.O.B 163
Mitcham, Victoria 3132   Tel. (03) 9210.7777
                          Fax: (03) 9210.7788

AUSTRIA – AUTRICHE
Gerold & Co.
Graben 31
Wien I   Tel. (0222) 533.50.14
          Fax: (0222) 512.47.31.29

BELGIUM – BELGIQUE
Jean De Lannoy
Avenue du Roi, Koningslaan 202
B-1060 Bruxelles   Tel. (02) 538.51.69/538.08.41
                    Fax: (02) 538.08.41

CANADA
Renouf Publishing Company Ltd.
5369 Canotek Road
Unit 1
Ottawa, Ont. K1J 9J3   Tel. (613) 745.2665
                        Fax: (613) 745.7660
Stores:
71 1/2 Sparks Street
Ottawa, Ont. K1P 5R1   Tel. (613) 238.8985
                        Fax: (613) 238.6041

12 Adelaide Street West
Toronto, QN M5H 1L6   Tel. (416) 363.3171
                       Fax: (416) 363.5963

Les Éditions La Liberté Inc.
3020 Chemin Sainte-Foy
Sainte-Foy, PQ G1X 3V6   Tel. (418) 658.3763
                          Fax: (418) 658.3763

Federal Publications Inc.
165 University Avenue, Suite 701
Toronto, ON M5H 3B8   Tel. (416) 860.1611
                       Fax: (416) 860.1608

Les Publications Fédérales
1185 Université
Montréal, QC H3B 3A7   Tel. (514) 954.1633
                        Fax: (514) 954.1635

CHINA – CHINE
Book Dept., China Natinal Publiations
Import and Export Corporation (CNPIEC)
16 Gongti E. Road, Chaoyang District
Beijing 100020   Tel. (10) 6506-6688 Ext. 8402
                       (10) 6506-3101

CHINESE TAIPEI – TAIPEI CHINOIS
Good Faith Worldwide Int'l. Co. Ltd.
9th Floor, No. 118, Sec. 2
Chung Hsiao E. Road
Taipei   Tel. (02) 391.7396/391.7397
          Fax: (02) 394.9176

CZECH REPUBLIC –
RÉPUBLIQUE TCHÈQUE
National Information Centre
NIS – prodejna
Konviktská 5
Praha 1 – 113 57   Tel. (02) 24.23.09.07
                    Fax: (02) 24.22.94.33
E-mail: nkposp@dec.niz.cz
Internet: http://www.nis.cz

DENMARK – DANEMARK
Munksgaard Book and Subscription Service
35, Nørre Søgade, P.O. Box 2148
DK-1016 København K   Tel. (33) 12.85.70
                       Fax: (33) 12.93.87

J. H. Schultz Information A/S,
Herstedvang 12,
DK – 2620 Albertslung   Tel. 43 63 23 00
                         Fax: 43 63 19 69
Internet: s-info@inet.uni-c.dk

EGYPT – ÉGYPTE
The Middle East Observer
41 Sherif Street
Cairo   Tel. (2) 392.6919
         Fax: (2) 360.6804

FINLAND – FINLANDE
Akateeminen Kirjakauppa
Keskuskatu 1, P.O. Box 128
00100 Helsinki

Subscription Services/Agence d'abonnements :
P.O. Box 23
00100 Helsinki   Tel. (358) 9.121.4403
                  Fax: (358) 9.121.4450

*FRANCE
OECD/OCDE
Mail Orders/Commandes par correspondance :
2, rue André-Pascal
75775 Paris Cedex 16   Tel. 33 (0)1.45.24.82.00
                        Fax: 33 (0)1.49.10.42.76
                        Telex: 640048 OCDE
Internet: Compte.PUBSINQ@oecd.org

Orders via Minitel, France only/
Commandes par Minitel, France exclusivement :
36 15 OCDE

OECD Bookshop/Librairie de l'OCDE :
33, rue Octave-Feuillet
75016 Paris   Tel. 33 (0)1.45.24.81.81
               33 (0)1.45.24.81.67

Dawson
B.P. 40
91121 Palaiseau Cedex   Tel. 01.89.10.47.00
                         Fax: 01.64.54.83.26

Documentation Française
29, quai Voltaire
75007 Paris   Tel. 01.40.15.70.00

Economica
49, rue Héricart
75015 Paris   Tel. 01.45.78.12.92
               Fax: 01.45.75.05.67

Gibert Jeune (Droit-Économie)
6, place Saint-Michel
75006 Paris   Tel. 01.43.25.91.19

Librairie du Commerce International
10, avenue d'Iéna
75016 Paris   Tel. 01.40.73.34.60

Librairie Dunod
Université Paris-Dauphine
Place du Maréchal-de-Lattre-de-Tassigny
75016 Paris   Tel. 01.44.05.40.13

Librairie Lavoisier
11, rue Lavoisier
75008 Paris   Tel. 01.42.65.39.95

Librairie des Sciences Politiques
30, rue Saint-Guillaume
75007 Paris   Tel. 01.45.48.36.02

P.U.F.
49, boulevard Saint-Michel
75005 Paris   Tel. 01.43.25.83.40

Librairie de l'Université
12a, rue Nazareth
13100 Aix-en-Provence   Tel. 04.42.26.18.08

Documentation Française
165, rue Garibaldi
69003 Lyon   Tel. 04.78.63.32.23

Librairie Decitre
29, place Bellecour
69002 Lyon   Tel. 04.72.40.54.54

Librairie Sauramps
Le Triangle
34967 Montpellier Cedex 2   Tel. 04.67.58.85.15
                             Fax: 04.67.58.27.36

A la Sorbonne Actual
23, rue de l'Hôtel-des-Postes
06000 Nice   Tel. 04.93.13.77.75
              Fax: 04.93.80.75.69

GERMANY – ALLEMAGNE
OECD Bonn Centre
August-Bebel-Allee 6
D-53175 Bonn   Tel. (0228) 959.120
                Fax: (0228) 959.12.17

GREECE – GRÈCE
Librairie Kauffmann
Stadiou 28
10564 Athens   Tel. (01) 32.55.321
                Fax: (01) 32.30.320

HONG-KONG
Swindon Book Co. Ltd.
Astoria Bldg. 3F
34 Ashley Road, Tsimshatsui
Kowloon, Hong Kong   Tel. 2376.2062
                      Fax: 2376.0685

HUNGARY – HONGRIE
Euro Info Service
Margitsziget, Európa Ház
1138 Budapest   Tel. (1) 111.60.61
                 Fax: (1) 302.50.35
E-mail: euroinfo@mail.matav.hu
Internet: http://www.euroinfo.hu//index.html

ICELAND – ISLANDE
Mál og Menning
Laugavegi 18, Pósthólf 392
121 Reykjavik   Tel. (1) 552.4240
                 Fax: (1) 562.3523

INDIA – INDE
Oxford Book and Stationery Co.
Scindia House
New Delhi 110001   Tel. (11) 331.5896/5308
                    Fax: (11) 332.2639
E-mail: oxford.publ@axcess.net.in

17 Park Street
Calcutta 700016   Tel. 240832

INDONESIA – INDONÉSIE
Pdii-Lipi
P.O. Box 4298
Jakarta 12042   Tel. (21) 573.34.67
                 Fax: (21) 573.34.67

IRELAND – IRLANDE
Government Supplies Agency
Publications Section
4/5 Harcourt Road
Dublin 2   Tel. 661.31.11
            Fax: 475.27.60

ISRAEL – ISRAËL
Praedicta
5 Shatner Street
P.O. Box 34030
Jerusalem 91430   Tel. (2) 652.84.90/1/2
                   Fax: (2) 652.84.93

R.O.Y. International
P.O. Box 13056
Tel Aviv 61130   Tel. (3) 546 1423
                  Fax: (3) 546 1442
E-mail: royil@netvision.net.il

Palestinian Authority/Middle East:
INDEX Information Services
P.O.B. 19502
Jerusalem   Tel. (2) 627.16.34
             Fax: (2) 627.12.19

ITALY – ITALIE
Libreria Commissionaria Sansoni
Via Duca di Calabria, 1/1
50125 Firenze   Tel. (055) 64.54.15
                 Fax: (055) 64.12.57
E-mail: licosa@ftbcc.it

Via Bartolini 29
20155 Milano   Tel. (02) 36.50.83

Editrice e Libreria Herder
Piazza Montecitorio 120
00186 Roma   Tel. 679.46.28
              Fax: 678.47.51

Libreria Hoepli
Via Hoepli 5
20121 Milano   Tel. (02) 86.54.46
                Fax: (02) 805.28.86

Libreria Scientifica
Dott. Lucio de Biasio 'Aeiou'
Via Coronelli, 6
20146 Milano   Tel. (02) 48.95.45.52
                                Fax: (02) 48.95.45.48

JAPAN – JAPON
OECD Tokyo Centre
Landic Akasaka Building
2-3-4 Akasaka, Minato-ku
Tokyo 107   Tel. (81.3) 3586.2016
                                Fax: (81.3) 3584.7929

KOREA – CORÉE
Kyobo Book Centre Co. Ltd.
P.O. Box 1658, Kwang Hwa Moon
Seoul   Tel. 730.78.91
                                Fax: 735.00.30

MALAYSIA – MALAISIE
University of Malaya Bookshop
University of Malaya
P.O. Box 1127, Jalan Pantai Baru
59700 Kuala Lumpur
Malaysia   Tel. 756.5000/756.5425
                                Fax: 756.3246

MEXICO – MEXIQUE
OECD Mexico Centre
Edificio INFOTEC
Av. San Fernando no. 37
Col. Toriello Guerra
Tlalpan C.P. 14050
Mexico D.F.   Tel. (525) 528.10.38
                                Fax: (525) 606.13.07
E-mail: ocde@rtn.net.mx

NETHERLANDS – PAYS-BAS
SDU Uitgeverij Plantijnstraat
Externe Fondsen
Postbus 20014
2500 EA's-Gravenhage   Tel. (070) 37.89.880
Voor bestellingen: Fax: (070) 34.75.778

Subscription Agency/   Agence d'abonnements :
SWETS & ZEITLINGER BV
Heereweg 347B
P.O. Box 830
2160 SZ Lisse   Tel. 252.435.111
                                Fax: 252.415.888

**NEW ZEALAND –
NOUVELLE-ZÉLANDE**
GPLegislation Services
P.O. Box 12418
Thorndon, Wellington   Tel. (04) 496.5655
                                Fax: (04) 496.5698

NORWAY – NORVÈGE
NIC INFO A/S
Ostensjoveien 18
P.O. Box 6512 Etterstad
0606 Oslo   Tel. (22) 97.45.00
                                Fax: (22) 97.45.45

PAKISTAN
Mirza Book Agency
65 Shahrah Quaid-E-Azam
Lahore 54000   Tel. (42) 735.36.01
                                Fax: (42) 576.37.14

PHILIPPINE – PHILIPPINES
International Booksource Center Inc.
Rm 179/920 Cityland 10 Condo Tower 2
HV dela Costa Ext cor Valero St.
Makati Metro Manila   Tel. (632) 817 9676
                                Fax: (632) 817 1741

POLAND – POLOGNE
Ars Polona
00-950 Warszawa
Krakowskie Prezdmiescie 7   Tel. (22) 264760
                                Fax: (22) 265334

PORTUGAL
Livraria Portugal
Rua do Carmo 70-74
Apart. 2681
1200 Lisboa   Tel. (01) 347.49.82/5
                                Fax: (01) 347.02.64

SINGAPORE – SINGAPOUR
Ashgate Publishing
Asia Pacific Pte. Ltd
Golden Wheel Building, 04-03
41, Kallang Pudding Road
Singapore 349316   Tel. 741.5166
                                Fax: 742.9356

SPAIN – ESPAGNE
Mundi-Prensa Libros S.A.
Castelló 37, Apartado 1223
Madrid 28001   Tel. (91) 431.33.99
                                Fax: (91) 575.39.98
E-mail: mundiprensa@tsai.es
Internet: http://www.mundiprensa.es

Mundi-Prensa Barcelona
Consell de Cent No. 391
08009 – Barcelona   Tel. (93) 488.34.92
                                Fax: (93) 487.76.59

Libreria de la Generalitat
Palau Moja
Rambla dels Estudis, 118
08002 – Barcelona
                     (Suscripciones) Tel. (93) 318.80.12
                     (Publicaciones) Tel. (93) 302.67.23
                                Fax: (93) 412.18.54

SRI LANKA
Centre for Policy Research
c/o Colombo Agencies Ltd.
No. 300-304, Galle Road
Colombo 3   Tel. (1) 574240, 573551-2
                                Fax: (1) 575394, 510711

SWEDEN – SUÈDE
CE Fritzes AB
S–106 47 Stockholm   Tel. (08) 690.90.90
                                Fax: (08) 20.50.21

For electronic publications only/
Publications électroniques seulement
STATISTICS SWEDEN
Informationsservice
S-115 81 Stockholm   Tel. 8 783 5066
                                Fax: 8 783 4045

Subscription Agency/Agence d'abonnements :
Wennergren-Williams Info AB
P.O. Box 1305
171 25 Solna   Tel. (08) 705.97.50
                                Fax: (08) 27.00.71

Liber distribution
Internatinal organizations
Fagerstagatan 21
S-163 52 Spanga

SWITZERLAND – SUISSE
Maditec S.A. (Books and Periodicals/Livres
et périodiques)
Chemin des Palettes 4
Case postale 266
1020 Renens VD 1   Tel. (021) 635.08.65
                                Fax: (021) 635.07.80

Librairie Payot S.A.
4, place Pépinet
CP 3212
1002 Lausanne   Tel. (021) 320.25.11
                                Fax: (021) 320.25.14

Librairie Unilivres
6, rue de Candolle
1205 Genève   Tel. (022) 320.26.23
                                Fax: (022) 329.73.18

Subscription Agency/Agence d'abonnements :
Dynapresse Marketing S.A.
38, avenue Vibert
1227 Carouge   Tel. (022) 308.08.70
                                Fax: (022) 308.07.99

See also – Voir aussi :
OECD Bonn Centre
August-Bebel-Allee 6
D-53175 Bonn (Germany)   Tel. (0228) 959.120
                                Fax: (0228) 959.12.17

THAILAND – THAÏLANDE
Suksit Siam Co. Ltd.
113, 115 Fuang Nakhon Rd.
Opp. Wat Rajbopith
Bangkok 10200   Tel. (662) 225.9531/2
                                Fax: (662) 222.5188

**TRINIDAD & TOBAGO, CARIBBEAN
TRINITÉ-ET-TOBAGO, CARAÏBES**
Systematics Studies Limited
9 Watts Street
Curepe
Trinadad & Tobago, W.I.   Tel. (1809) 645.3475
                                Fax: (1809) 662.5654
E-mail: tobe@trinidad.net

TUNISIA – TUNISIE
Grande Librairie Spécialisée
Fendri Ali
Avenue Haffouz Imm El-Intilaka
Bloc B 1 Sfax 3000   Tel. (216-4) 296 855
                                Fax: (216-4) 298.270

TURKEY – TURQUIE
Kültür Yayinlari Is-Türk Ltd.
Atatürk Bulvari No. 191/Kat 13
06684 Kavaklidere/Ankara
                          Tel. (312) 428.11.40 Ext. 2458
                                Fax : (312) 417.24.90

Dolmabahce Cad. No. 29
Besiktas/Istanbul   Tel. (212) 260 7188

UNITED KINGDOM – ROYAUME-UNI
The Stationery Office Ltd.
Postal orders only:
P.O. Box 276, London SW8 5DT
Gen. enquiries   Tel. (171) 873 0011
                                Fax: (171) 873 8463

The Stationery Office Ltd.
Postal orders only:
49 High Holborn, London WC1V 6HB
Branches at: Belfast, Birmingham, Bristol,
Edinburgh, Manchester

UNITED STATES – ÉTATS-UNIS
OECD Washington Center
2001 L Street N.W., Suite 650
Washington, D.C. 20036-4922 Tel. (202) 785.6323
                                Fax: (202) 785.0350
Internet: washcont@oecd.org

Subscriptions to OECD periodicals may also be
placed through main subscription agencies.

Les abonnements aux publications périodiques de
l'OCDE peuvent être souscrits auprès des
principales agences d'abonnement.

Orders and inquiries from countries where Distribu-
tors have not yet been appointed should be sent to:
OECD Publications, 2, rue André-Pascal, 75775
Paris Cedex 16, France.

Les commandes provenant de pays où l'OCDE n'a
pas encore désigné de distributeur peuvent être
adressées aux Éditions de l'OCDE, 2, rue André-
Pascal, 75775 Paris Cedex 16, France.

                                             12-1996

OECD PUBLICATIONS, 9, rue de la Fédération, 75739 PARIS CEDEX 15
PRINTED IN FRANCE BY LOUIS-JEAN
(61 98 23 1 P) ISBN 92-64-16186-4 – 1998